CHEMICAL PROPERTIES AND ANALYSIS OF REFRACTORY COMPOUNDS

KHIMICHESKIE SVOISTVA I METODY ANALIZA TUGOPLAVKIKH SOEDINENII

ХИМИЧЕСКИЕ СВОЙСТВА И МЕТОДЫ АНАЛИЗА ТУГОПЛАВКИХ СОЕДИНЕНИЙ

CHEMICAL PROPERTIES AND ANALYSIS OF REFRACTORY COMPOUNDS

Edited by G. V. Samsonov

*Director, Laboratory of Metallurgy of Rare Metals and
Refractory Compounds
Institute of Powder Metallurgy and Special Alloys
Academy of Sciences of the Ukrainian SSR, Kiev, USSR*

With the cooperation of I. N. Frantsevich, V. N. Eremenko,
T. N. Nazarchuk, and O. I. Popova

Translated from Russian by G. D. Archard

Ⓒⓑ CONSULTANTS BUREAU · NEW YORK – LONDON · 1972

The original Russian text, published by Naukova Dumka in Kiev in 1969, has
been corrected by the editor for the present edition. The translation is
published under an agreement with Mezhdunarodnaya Kniga, the Soviet book
export agency.

Г. В. Самсонов

Library of Congress Catalog Card Number 70-37619
ISBN-13: 978-1-4615-8557-2   e-ISBN-13: 978-1-4615-8555-8
DOI: 10.1007/978-1-4615-8555-8

© 1972 Consultants Bureau, New York
Softcover reprint of the hardcover 1st edition 1972
A Division of Plenum Publishing Corporation
227 West 17th Street, New York, N. Y. 10011

United Kingdom edition published by Consultants Bureau, London
A Division of Plenum Publishing Company, Ltd.
Davis House (4th Floor), 8 Scrubs Lane, Harlesden, NW10 6SE, London, England

PREFACE

This collection sets out the results of various investigations into the chemical properties of refractory compounds and refractory-base alloys used in various fields of modern technology, together with original methods of analysis.

The book is intended for analytical chemists, engineers, workers in scientific-research establishments and industrial laboratories, graduates, and students of the senior courses in chemical and metallurgical higher-education institutions.

CONTENTS

Interaction of the Carbides of Group IV and V Transition Metals with Various Acids
E. E. Kotlyar and T. N. Nazarchuk 1

Method of Quantitative X-ray Analysis for Determining the Amount of Free Carbon in
Boron Carbide
M. I. Sokhor and G. V. Sofronov 6

Method of Separating and Determining the Free Carbon in Materials Containing
Refractory Compounds
L. A. Mashkovich and A. F. Kuteinikov 14

Stability of Boron–Carbon Compounds in Oxygen at High Temperatures
L. E. Pechentkovskaya and T. N. Nazarchuk 20

Certain Chemical Properties of Boron Carbonitride
L. E. Pechentkovskaya and T. N. Nazarchuk 25

Oxidation of Boron, Gallium, and Indium Phosphides in Air
L. L. Vereikina 29

High-Temperature Oxidation Resistance of Refractory Silicon Nitride–Silicon Carbide
Materials
I. N. Godovannaya and O. I. Popova 33

Production and Chemical Stability of the Hydrides of Group IV and V Transition Metals
M. M. Antonova 36

Chemical Analysis of the Reaction Products of Boron with Arsenic and Phosphorus
A. A. Reshchikova and Z. S. Medvedeva 42

Complexonometric Analysis of Molybdenum Alloys
L. N. Kugai, O. F. Galadzhii, and V. I. Kornilova 48

Analysis of Titanium, Zirconium, Hafnium, and Tantalum Germanides
G. T. Kabannik and O. I. Popova 54

Some Data Relating to the Chemical Properties of Germanides
O. I. Popova 57

Analysis of the Selenides of the Rare-Earth Elements
V. A. Obolonchik and T. M. Mikhlina 61

Complexonometric Analysis of Alloys of the Rare-Earth Oxides with the Oxides of Group
II Elements and Chromium
S. F. Boremskaya and G. T. Kabannik 64

Separation of Alloyed Chromium Carbides of the Cr_2C (Metastable) and $Cr_{23}C_6$
or Cr_7C_3 Types Isolated from Steels and Alloys
 L. V. Zaslavskaya, N. V. Ivanova, and N. F. Lashko 68

Chemical and Electrochemical Methods of Separating the Carbides MeC and the
Carbonitrides Me(C, N) of Group IV and V Metals
 G. G. Georgieva, N. F. Lashko, and K. P. Sorokina 72

Electrolytic Isolation of the σ Phase from Heat-Resistant Steels and the Determination
of Its Composition
 E. F. Yakovleva, I. M. Dubrovina, and L. V. Stegnukhina 77

Interaction of Refractory Group V Metals with Zinc
 A. P. Obukov, V. N. Gurin, I. R. Kozlova, Z. P. Terent'eva, and T. I. Mazina 81

Determination of Tungsten in Binary Tungsten–Molybdenum Alloys
 Z. S. Mukhina, L. I. Il'ina, and N. S. Kondukova 86

Determination of Molybdenum in the Presence of Tungsten
 V. G. Shcherbakov, Z. K. Stegendo, and R. A. Antonova 91

Photometric Determination of Boron in Nickel and Titanium Borides Using Magnezone I
in an Alkaline Medium
 E. I. Nikitina 94

Spectrophotometric Study of the Formation of Compounds between Niobium and the
Reagent PAN
 V. I. Kornilova 100

Determination of Arsenic in High-Purity Molybdenum
 V. G. Shcherbakov and G. V. Onuchina 103

Chemical Phase Analysis of Mixtures of Borides, Carbides, and Borocarbides
 N.V. Vekshina and L. Ya. Markovskii 106

Interaction of Borides with Carbon and Carbides
 L. Ya. Markovskii, N. V. Vekshina, and E. T. Bezruk 114

Chemical Properties and Analysis of Certain Transition Metal Sulfides
 S. V. Radzikovskaya and V. F. Bukhanevich 119

Analysis of the Silicides of Group IV to VI Transition Elements
 V. P. Kopylova and T. N. Nazarchuk 123

INTERACTION OF THE CARBIDES OF GROUP IV AND V TRANSITION METALS WITH VARIOUS ACIDS

E. E. Kotlyar and T. N. Nazarchuk

Institute of Problems in Materials Science, Academy of Sciences of the Ukrainian SSR

The interaction of the carbides of Group IV and V transition metals with acids and alkalis has hitherto mainly been studied in order to secure qualitative or semiquantitative data relating to their chemical stability; no attention has been paid to the composition of the gaseous decomposition products [2].

The aim of the present investigation was to study the interaction of carbides with phosphoric, sulfuric, and nitric acids, to establish the composition of the gaseous decomposition products, and to use the results in discussing the nature of the chemical bond in the carbides.

The interaction of the carbides with concentrated phosphoric, sulfuric, and nitric acids was studied in an inert-gas (nitrogen) atmosphere, heating being carried out in the apparatus illustrated in Fig. 1.

A 0.1-0.3 g sample of carbide was placed in a Wurtz flask, into which 30-40 ml of acid were poured. In order to study the interaction of the carbides with the sulfuric acid, 50 ml of 80% isopropyl alcohol solution were poured into the first washing bottle to absorb the SO_3 [4] and 50 ml of 2% iodine solution into the second to absorb the SO_2, while in studying the interaction of the carbides with nitric acid both washing bottles were filled with 1:1 sulfuric acid in order to remove the nitrogen oxides. Then the whole system was connected as indicated in Fig. 1.

In order to displace the air before the beginning of the reaction, nitrogen was passed slowly through the whole system for 2-2.5 h (3-4 bubbles per second), after which the reaction mixture was heated (with the tap open in the gas burette) and the gases evolved were collected. When the reaction had finished, the solution was analyzed for metal content, while the undissolved residue was analyzed for carbide and free carbon. The gas mixture was analyzed chromatographically. The carbon material balance was determined to an accuracy of 3-5%.

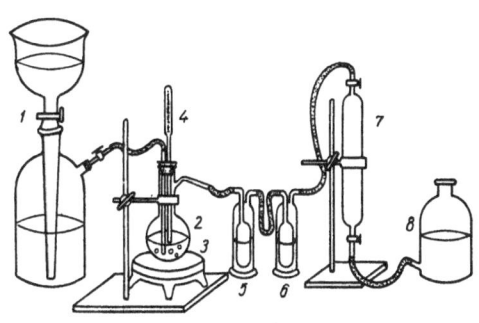

Fig. 1. Apparatus for studying the interaction of carbides with acids: 1) Gas-holder with nitrogen; 2) Wurtz flask; 3) electric plate; 4) thermometer; 5) bottle containing 80% isopropyl alcohol for absorbing SO_3; 6) bottle containing 2% iodine solution for absorbing SO_2; 7) gas burette; 8) pressure-head bottle.

TABLE 1. Composition of the Products of the Interactions between Group IV and V Transition Metal Carbides and Sulfuric, Phosphoric, and Nitric Acids

Products of the interaction	Reaction temperature, °C	Reaction time, h	Decomposing carbide, %	Amorphous carbon, %	Composition of gas mixture (chromatographic method)			
					CH_4	CO	CO_2	H_2
$ZrC + H_3PO_4$ ($d = 1.7$ g/cm^3)	260—270	1	100.0	—	91.4	—	7.8	0.8
	250—255	1	100.0	—	87.3	—	12.0	0.7
$HfC + H_3PO_4$ ($d = 1.7$ g/cm^3)	230—240	1.5	77.0	—	62.1	—	35.2	3.0
	250	1	100.0	—	77.2	—	19.3	3.4
	250—270	1.5	100.0	—	69.3	—	28.9	1.7
$TiC + H_2SO_4$ ($d = 1.84$ g/cm^3)	240—250	2	74.5	57.0	—	18.0	8.0	2.0
	250—270	2	76.0	33.0	—	6.3	92.4	1.3
	315—320	1	100.0	25.7	—	5.5	92.0	1.5
$ZrC + H_2SO_4$ ($d = 1.84$ g/cm^3)	260	1	100.0	14.2	48.2	—	48.2	3.0
	200—235	1.5	100.0	7.0	71.0	—	26.5	2.5
$HfC + H_2SO_4$ ($d = 1.84$ g/cm^3)	264—268	1	100.0	16.1	64.2	—	31.8	4.0
	290—300	1	100.0	6.3	16.2	—	81.3	2.4
$VC + H_2SO_4$ ($d = 1.84$ g/cm^3)	260	1.5	—	—	—	20.0	78.0	2.0
	300	1	—	—	—	—	96.4	3.6
$NbC + H_2SO_4$	260	1.5—2	93.0	36.3	—	19.7	79.0	1.3
	280	1.5	92.0	47.0	—	24.3	73.8	1.9
	290—310	1	100.0	—	—	13.7	82.9	3.4
	265	1.5	81.0	13.0	—	5.4	92.3	2.3
$TaC + H_2SO_4$	255	1.5	75.0	10.0	—	2.2	96.0	1.6
	276	1.0	80.0	6.0	—	6.0	92.3	1.7
$ZrC + HNO_3$	110—120	0.5	100.0	—	Trace	—	95—97	2—3
$HfC + HNO_3$	110—120	0.5	100.0	—	Trace	—	95—97	2—3
$NbC + HNO_3 + NH_4F$	110—120	0.5	100.0	—	—	Trace	95—97	2—3
$TaC + HNO_3 + NH_4F$	110—120	0.5	100.0	—	—	Trace	95—97	2—3

At 230-260°C, concentrated phosphoric acid (d = 1.7) only decomposes zirconium and hafnium carbides. Decomposition starts at 230-240°C, and at 250° the dissolution proceeds very energetically; on boiling for 30–40 min these carbides decompose completely. Titanium, vanadium, niobium, and tantalum carbides are unaffected by phosphoric acid, even on heating to 300°C.

The main gaseous decomposition product resulting from the interaction of hafnium and zirconium carbides with phosphoric acid is methane (70-90 vol.%). Apart from this, the products include hydrogen (1-3%) and CO_2 (Table 1).

The resultant solutions are dense and transparent; on heavily diluting with water, amorphous white precipitates appear, and, after heating, these transform into pyrophosphates $Zr_2P_2O_7$ and $Hf_2P_2O_7$. According to Brauer [5], the acid zirconium and hafnium phosphates (biphosphates) $ZrO(H_2PO_4)_2$, $HfO(H_2PO_4)_2$ in the freshly precipitated state dissolve in a mixture of phosphoric, oxalic, and sulfuric acids with the formation of soluble complexes. Evidently these complexes are also formed in the presence of a large excess of concentrated phosphoric acid. Diluted with water, i.e., on reducing the concentration of phosphoric acid, they transform into poorly-soluble middle phosphates and are precipitated.

The decomposition of the hafnium and zirconium carbides by phosphoric acid may be represented by the equations

$$ZrC + 2H_3PO_4 + H_2O = ZrO(H_2PO_4)_2 + CH_4;$$
$$HfC + 2H_3PO_4 + H_2O = HfO(H_2PO_4)_2 + CH_4.$$

Concentrated sulfuric acid decomposes all the carbides between 200 and 300°C. The decomposition of hafnium and zirconium carbides starts at 200-220°C; at 230-250°C they are dissolved very energetically, and decompose completely in 30-40 min. The titanium, niobium, and tantalum carbides decompose by 70-80% on boiling for 1.5 h at 250-260°C, and decompose completely after boiling for 1.5-2 h at 290-310°C. The decomposition of all the carbides is accompanied by the release of amorphous carbon, which gives the solution a chestnut color in the finely-dispersed state. At the beginning of the interaction the solution and the walls of the vessel show amorphous carbon flakes; at the end of the reaction the solution clears and at 290-310°C it becomes transparent.

The precipitation of amorphous carbon was observed by Greenwood and Osborne [6] when lanthanum dicarbide interacted with 4 N sulfuric acid. The solution of 2% iodine in the washing bottle becomes much lighter during the reaction; frequently it is completely decolored by the evolving SO_2.

It follows from the tabulated data that the interaction of zirconium and hafnium carbides with sulfuric acid differs from that of titanium, vanadium, niobium, and tantalum carbides, not only in respect of the temperature range of decomposition and the period of dissolution, but also in respect of the composition of the gaseous products. The gaseous products of the decomposition of zirconium and hafnium carbides contain methane (50-70 vol.%), CO_2 (30-50 vol.%), and hydrogen (2.5-4%). The ratio of methane, CO_2, and amorphous carbon depends on the temperature.

The principal reaction governing the decomposition of hafnium and zirconium carbides by sulfuric acid may be expressed as follows:

$$HfC + 2H_2SO_4 = Hf(SO_4)_2 + CH_4,$$
$$HfC + 4H_2SO_4 = Hf(SO_4)_2 + C + 2SO_2 + 4H_2O,$$
$$ZrC + 2H_2SO_4 = Zr(SO_4)_2 + CH_4,$$
$$ZrC + 4H_2SO_4 = Zr(SO_4)_2 + 2SO_2 + C + 4H_2O.$$

Secondary reactions:

$$C + 2H_2SO_4 = CO_2 + 2H_2O + 2SO_2,$$
$$CH_4 + 4H_2SO_4 = CO_2 + 6H_2O + 4SO_2.$$

The interaction of titanium, vanadium, niobium, and tantalum carbides with concentrated sulfuric acid results in the precipitation of amorphous carbon. The amount of amorphous carbon relative to the whole of the decomposing carbon in the titanium carbide is 57-33% between 240 and 270°C; in the case of niobium carbide the figures are 47-36%. Vanadium carbide decomposes partly, with the formation of vanadium sulphate (green solution) and a poorly soluble yellow precipitate, which, according to analysis, has the formula composition $V_2(SO_4)_3$; in all other cases the solutions are viscous and transparent after decomposition, and no precipitates are formed on diluting with water. Hydrolysis of the solutions only occurs after prolonged standing. It thus follows that titanium, niobium, and tantalum carbides decompose with the formation of oxysulphates stable in concentrated H_2SO_4 [1].

The composition of the gaseous products includes carbon monoxide (5-25%), carbon dioxide (70-90%), and hydrogen (1.5-3.0%). No methane is found in the gas phase at any temperature.

The ratio of the amorphous carbon, carbon monoxide, and carbon dioxide depends on the temperature. This would suggest that the CO and CO_2 are not the direct products of the decomposition of the carbides but are formed as a result of the oxidation of the amorphous carbon by the concentrated sulfuric acid.

The decomposition of the titanium, niobium, and tantalum carbides by fuming sulfuric acid is represented by the equations:

$$TiC + H_2SO_4 + SO_3 = TiOSO_4 + C + SO_2 + H_2O,$$
$$VC + 6H_2SO_4 = V_2(SO_4)_3 + 2C + 3SO_2 + 6H_2O,$$
$$2NbC + 4H_2SO_4 + 5SO_3 = Nb_2O(SO_4)_4 + 2C + 5SO_2 + 4H_2O.$$

Nitric acid (d = 1.43) and nitric acid (1:1) decompose zirconium, hafnium, titanium, and vanadium carbides at 100-110°C. No carbon is precipitated in the decomposition; after the dissolution of the vanadium, zirconium, and hafnium carbides, the solutions are transparent, i.e., the interaction involves the formation of soluble nitrates. Titanium carbide is decomposed completely with the precipitation of titanium hydroxide $TiO_2 \cdot nH_2O$.

In the decomposition of zirconium and hafnium carbides, the gas fraction incorporates 95-97% of CO_2, 2-3% of hydrogen, and traces of methane. In the decomposition of vanadium and titanium carbides, the gas products contain hydrogen (2-3%) and traces of carbon monoxide as well as CO_2.

Niobium and tantalum carbides are decomposed by nitric acid at 100-120°C in the presence of a 5% solution of ammonium fluoride. After decomposition the solutions are transparent, no precipitation of amorphous carbon taking place. The gaseous products include carbon dioxide (95-97%), hydrogen (2-3%), and traces of carbon monoxide.

The interaction of the carbides with nitric acid may be represented by the equations:

$$ZrC + 6HNO_3 = ZrO(NO_3)_2 + CO_2 + 2NO + 2NO_2 + 3H_2O,$$
$$HfC + 6HNO_3 = HfO(NO_3)_2 + CO_2 + 2NO + 2NO_2 + 3H_2O,$$
$$VC + 4HNO_3 = VONO_3 + CO_2 + NO_2 + 2NO + 2H_2O,$$
$$TiC + 4HNO_3 = TiO_2 \cdot 2H_2O + CO_2 + 2NO + 2NO_2.$$

It follows from the foregoing data that only zirconium and hafnium carbides are decomposed by phosphoric and sulfuric acids with the evolution of methane (as the principal gaseous product), i.e., the carbon in these carbides is present in the form of C^4 groups possessing the stable sp^3 electron configuration. The even bonds of the carbon are symmetrical and identical; they form covalent polarized bonds with the metal, and this leads to considerable polarization of the ZrC and HfC molecules, although clearly the ionic bond is not the predominant one in these carbides, since they are not decomposed by water and dilute acids like carbides of the ionic type such as ScC, YC, and others. The absence of methane from the composition of the gaseous products in the decomposition of titanium, vanadium, niobium, and tantalum carbides indicates that the ionic Me−C bond is either entirely absent from these carbides or at any rate contribute very little.

As we have found, the carbides of the group IV transition metals (particularly zirconium and hafnium) are chemically much less stable than the group V carbides; the chemical stability of the group IV carbides diminishes with increasing atomic number of the transition metal (from titanium carbide to hafnium carbide), and this may be explained, as indicated in the foregoing, by the increase in the proportion of ionic bond in the same order. The chemical stability of the group V carbides, on the other hand, increases with increasing atomic number, from vanadium carbide to tantalum carbide.

In respect of its chemical properties, tantalum carbide resembles the carbides of the covalent type. This is evidently due to the fact that, in the carbides of group V metals, the covalent bonds between the atoms of the transition metals are particularly sharply expressed; the proportion of the covalent Me−Me bond increases with diminishing acceptor capacity of the metal, from vanadium carbide to tantalum carbide [3].

Conclusions

Concentrated phosphoric acid only decomposes zirconium and hafnium carbides at 230-250°C with the evolution of methane (70-90 vol.%). Titanium, vanadium, niobium, and tantalum carbides are not decomposed by phsophoric acid, even at 300°C.

Concentrated sulfuric acid decomposes all the carbides between 200 and 300°C. Zirconium and hafnium carbide decompose at 200-230°C with the formation of amorphous carbon and the evolution of methane as the principal gaseous product (50-70 vol.%). Titanium, vanadium, niobium, and tantalum carbides decompose at 250-300°C with the formation of amorphous carbon. The gaseous products include hydrogen, CO, and CO_2. Nitric acid, as a strong oxidizing agent, decomposes titanium, niobium, zirconium, and vanadium carbides with the evolution of carbon dioxide. Niobium and tantalum carbides are decomposed by nitric acid in the presence of ammonium fluoride.

The results of our analysis of the gaseous products obtained in the decomposition of the carbides by phosphoric and sulfuric acids confirm the corresponding chemical-stability data and lead to the conclusion that zirconium and hafnium carbides possess a considerable proportion of an ionic Me−C bond.

Literature Cited

1. Ya. Goroshchenko, Chemistry of Niobium and Tantalum [in Russian], Izd. AN Ukr.SSR, Kiev (1965), p. 349.
2. V. P. Kopylova, Zh. Prikladnoi Khim., 34:1936-1937 (1961).
3. G. V. Samsonov, Poroshkovaya Met., No. 1, p. 99 (1965).
4. A. V. Stepanov and A. V. Puchkin, Teploénergetika, 8:50 (1961).
5. G. Brauer, Handbuch präp. anorg. Chem., 3:934 (1954).
6. N. N.Greenwood and A. J. Osborn, J. Chem. Soc., 4:1775 (1961).

METHOD OF QUANTITATIVE X-RAY ANALYSIS FOR DETERMINING THE AMOUNT OF FREE CARBON IN BORON CARBIDE

M. I. Sokhor and G. V. Sofronov

All-Union Scientific-Research Institute of Analytical Processes

The method principally employed at the present time for determining the amount of free carbon in boron carbide is chemical analysis. There are two well-known chemical methods: the direct determination of C_{free}, and an indirect determination based on the overall composition (B_{tot} and C_{tot}). The direct chemical analysis for C_{free} is based on treating the boron carbide with a chromic mixture and plotting a graph of the amount of CO_2 evolved against the period of oxidation. However, in treating the sample with the chromic mixture partial oxidation of the boron carbide itself occurs, and the results of the chemical analysis for C_{free} may therefore be too high. Furthermore, it is assumed in both chemical methods that the boron carbide is always of the same composition given by the formula B_4C, while the C_{free} is the proportion of the carbon actually free [5]. The assumption as to the constant composition of the carbon and boron carbide leads to a second assumption, namely, that the properties of the two phases, and in particular their oxidizability, remain constant.

However, there is no certain evidence as to the constancy of the properties of the carbon and boron carbide in different samples. Earlier investigations (1957-1961) showed that the free carbon in boron carbide was frequently not pure graphite, but that it contained boron in the form of a solid solution of varying concentration; boron carbide had a wide range of compositions, from B_4C to $B_{6.75}C$ [3, 8].

These doubtful assumptions underlying the existing chemical methods of quantitatively determining the amount of free carbon C_{free} in boron carbide have made it desirable to develop an x-ray method of quantitative determination, allowing for the composition and structural state of the boron carbide and the carbon phase [9].

Figure 1 presents the Debye photographs of single-phase samples of boron carbide and graphite and a sample of commercial boron carbide containing free carbon. The x-ray method under consideration is based on determining the intensity of the (002) line of graphite relative to the intensity of the (110) line of boron carbide. The choice of these lines is due to the fact that the amount of free carbon in the commercial boron carbide is usually no greater than a few percent, so that the graphite phase appearing on the x-ray diffraction patterns of the commercial boron carbide will usually only be represented by the single, very strong graphite (002) line, with $\Theta_{Co} \approx 15°$, the intensity of this line on the x-ray diffraction patterns being similar to that of the neighboring (110) line of the boron carbide phase with $\Theta_{Co} \approx 18.5°$ (Fig. 2).

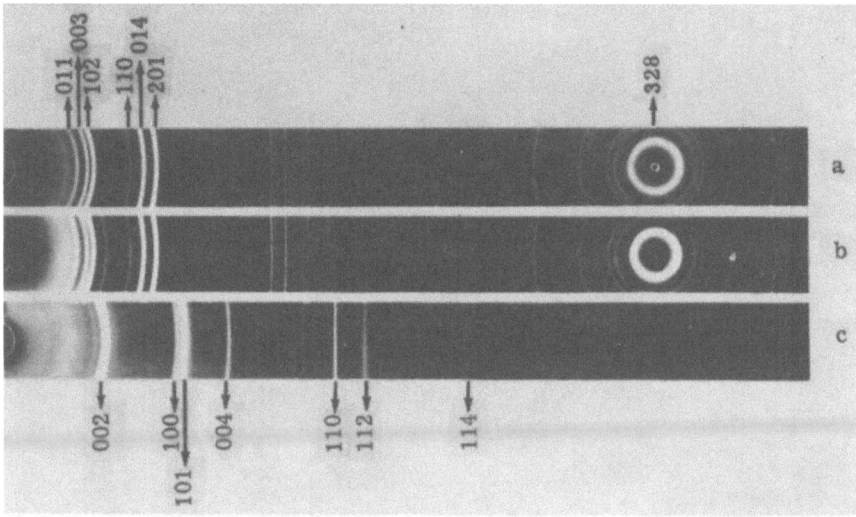

Fig. 1. Debye photographs of single-phase boron carbide $B_{4.78}C$ (a), commercial boron carbide containing the nominal compound together with 2% of free carbon (b), and ARV graphite (c). Co K_α radiation.

We developed the x-ray quantitative analysis of the C_{free} content on the basis of Palatnik's superposition technique [4].

The superposition method is based on comparing the x-ray diffraction patterns of the systems under analysis with standard superposition x-ray diffraction patterns obtained by x-ray photographing the single-phase constituents of the two-phase system on a single film. By varying the ratio of the exposures of the individual components we may vary the ratio of the line intensities on the superposition photographs, which is equivalent to varying the concentration of the components in the mixture.

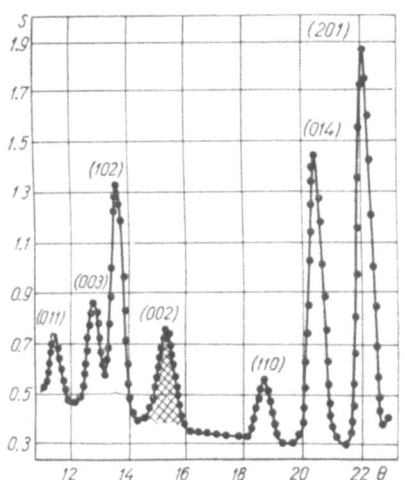

Fig. 2. Microphotometer recording of the front section of the Debye photograph of commercial boron carbide containing free carbon. (Six forward lines of boron carbide and the (002) line of graphite are illustrated.) Co K_α radiation.

In order to obtain the superposition Debye photographs, a special Debye camera (D-85) 85 mm in diameter had to be developed and constructed so as to facilitate the successive photographing of two or more samples on the same film without unloading the camera [6]. The form of the camera (with an indication of the individual units) is shown in Fig. 3. In contrast to the usual Debye cameras, the D-85 has interchangeable holding units for the cylindrical sample, enabling the sample to be adjusted outside the camera by means of a special adjusting device (Fig. 4). The holder with its duly adjusted sample is placed in the already loaded camera. The camera records lines corresponding to reflection angles Θ between 7 and 87.5° [2]. The sample is provided with rotation during the exposure.

Different carbon-containing materials are characterized by fine differences in the degree of perfection of their crystal lattices and the shape and size of the graphite crystals. These differences are reflected in the Debye photographs (Fig. 5). A particular feature is the fact that the structural state of the carbon phase changes the fine

Fig. 3. The D-85 camera in dismantled form: 1) Body; 2) support
connecting the body to the base of the camera; 3) sliding bearing with
a conical socket for the sample holder; 4) pulley attached to bearing 3;
5) conical socket with the sample holder unit installed; 6) conical rod
of the sample holder; 7) permanent magnet press-fitted to the rod 6;
8) soft iron disc with sample attached (the sample is adjusted by mov-
ing the disc with the magnet 7); 9) insertion piece guiding the collimator 10; 11) needle entering into the collimator 10 and cutting out
a beam of x rays 0.5-0.8 mm in diameter [7]; 12) nut for setting the
filter and fixing the trap (this is screwed on to the insertion piece 9);
13) primary-beam trap; 14) insertion piece for siting the trap 13;
15) nut with fluorescent screen (screwed on to the insertion piece 14).

structure and absolute intensity of the strongest line, the (002), which is used to determine the
C_{free} content of the sample. Figure 6 shows the dependence of the (002) line intensity of petroleum coke on the temperature of graphitization as obtained experimentally. The intensity of
the (002) line of ARV graphite is arbitrarily taken as 100%. We see that the intensity of the
(002) line of petroleum coke increases with increasing graphitization temperature. However,
the x-ray pictures show that the diffuseness of the edges of the (002) line remains unaltered.
For graphites of the ARV or Acheson type the (002) line is sharp and clear with a high intensity and no splitting. Natural graphite gives a split (002) line due to the texture of the sample.

Many years of experience in microscope and x-ray diffraction work carried out in our
Institute (by V. G. Kondakov and A. A. Kalinina) in relation to various types of commercial
boron carbide have shown that the structural state of the carbon phase in boron carbide may
vary considerably, being determined by the manufacturing conditions. In high-quality boron
carbide the free carbon occurs in the form of a finely-dispersed phase, often forming a eutec-
tic with the boron carbide and constituting a solid solution of boron in graphite. In relation to
the x-ray effect, this kind of finely-dispersed graphite in boron carbide may be formally likened
to carbon material with a low degree of graphitization, for example, petroleum coke roasted at
1550°C. In products obtained from melts of commercial boron carbide, and very frequently in

Fig. 4. D-85 camera in assembled form and adjusting attachment enabling the sample to be adjusted outside the camera.

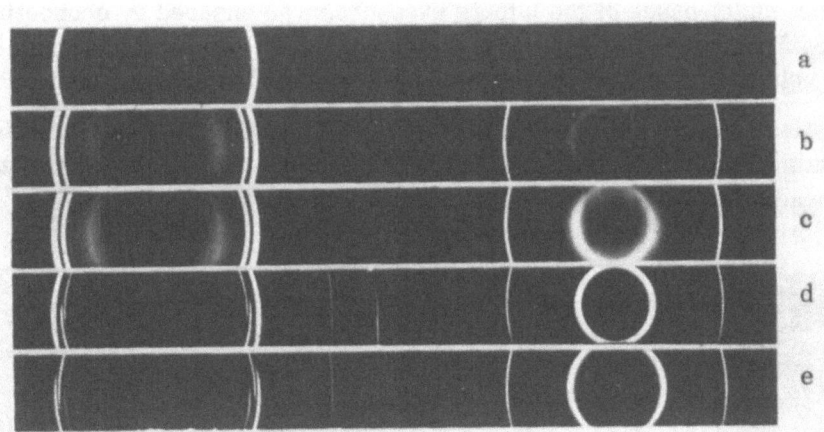

Fig. 5. Debye photographs of various carbon-containing materials.
a) Petroleum coke roasted at 1550° for 1.5 h; b) Acheson graphite;
c) ARV graphite; d) natural graphite; e) graphite extracted from
contaminated boron carbide. Cr radiation.

pressed boron–carbon alloys, we encounter graphite in the form of larger precipitates also.
This graphite gives an x-ray diffraction pattern similar to that of ARV or Acheson graphite.
In some cases of boron–carbon alloys prepared by repeated hot pressing, or in graphitized
fused boron carbide in which the boron carbide decomposes, the graphite separates in the form
of perfect petal-like crystals, giving a texture during the preparation of the sample. The x-ray
diffraction patterns of this kind of graphite are similar to those of natural graphite as regards
the relative intensity and fine structure of the lines (Fig. 5e), although an increase in the lattice
constants a and (particularly) c relative to those of natural graphite is often indicated, this

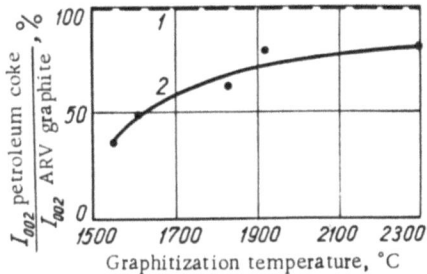

Fig. 6. Intensity of the (002) line
of petroleum coke as a function of
graphitization temperature: 1)
ARV graphite; 2) petroleum coke.

being due to the dissolution of boron in the graphite separated during the decomposition of the boron carbide.

In order to obtain a series of superposition Debye photographs, we used original components consisting of ARV graphite (0.05% moisture, 0.16% ash, 0.95% volatile materials) and boron carbide corresponding to the composition $B_{4.78}C$, not containing any free carbon (according to chemical analysis) and only giving the lines of boron carbide on the x-ray picture.

The foregoing differences in the structural state of the carbon phase in different samples of commercial boron carbide necessitated the photographing of a further series of superposition x-ray diffraction patterns with other samples of carbon-containing materials as components. Thus in addition to ARV graphite we used natural graphite and petroleum coke roasted at 1550°C. In order to allow for the influence of the composition of the boron carbide on the intensity of its lines, we studied $B_{6.75}C$ as well as $B_{4.78}C$.

Figure 7 shows the forward sections of the superposition Debye photographs of boron carbide $B_{4.78}C$ and ARV graphite taken with different exposure ratios of the components. The exposure time for the boron carbide samples in the superposition Debye photographs was as follows: for $B_{4.78}C$ 2.5 h and for $B_{6.75}C$ 5 h. The exposure time for the carbon component varied from 1 to 60 min. The exposure was maintained by reference to the clock in the URS-70 x-ray equipment. Exact maintenance of the minute exposures was ensured by connecting an RV time relay to the cut-off circuit of the high-voltage supply of the URS-70; this automatically disconnected the high voltage when the exposure time had expired, the accuracy being ± 0.05 min.

The operating conditions of the URS-70 were kept constant by stabilizing the voltage on the primary winding of the high-voltage transformer by means of an SN-1 stabilizer (127 ± 0.5 V). We used Co K_α radiation.

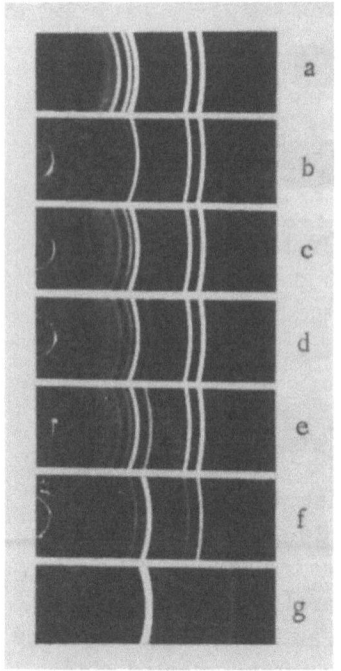

Fig. 7. Forward part of the superposition Debye photographs of boron carbide $B_{4.78}C$ and ARV graphite taken with different exposure ratios of the components. a) Boron carbide $B_{4.78}C$; b-f) boron carbide + graphite; g) graphite. Exposure ratio graphite/boron carbide: b) 0.03; c) 0.05; d) 0.08; e) 0.13; f) 0.5. Co K_α radiation.

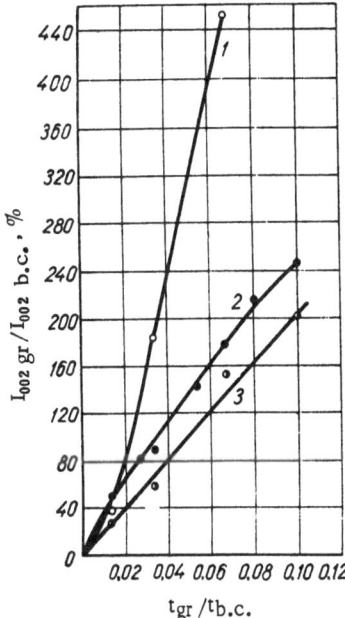

Fig. 8. Calibration curves plotted from series of superposition Debye photographs of various boron carbides and carbon materials: 1) $B_{6.75}C$ + ARV graphite; 2) $B_{4.78}C$ + ARV graphite; 3) $B_{4.78}C$ + petroleum coke roasted at 1550°C.

In practice, the quantitative determination of the C_{free} in boron carbide from the superposition Debye photographs amounted to the following. First a series of superposition Debye photographs of boron carbide and graphite was taken in the D-85 camera with various exposure ratios $t_{gr}/t_{b.c.}$; for each photograph the intensity ratio of the graphite and boron carbide lines $I_{002\,gr}/I_{110\,b.c.}$ was determined, and a calibration curve $I_{002\,gr}/I_{110\,b.c.} = f(t_{gr}/t_{b.c.})$ was plotted.

Allowing for the multitude of possible structural states of C_{free} in the boron carbide and also the wide range of possible phase compositions (from B_4C to $B_{6.75}C$), it is also essential to construct calibration curves for a number of combinations of boron carbides of different composition with different carbon-containing materials. Then the sample to be analyzed is photographed in the D-85 or some other Debye camera. By examining the fine structure (profile, width, intensity distribution) of the (002) line of the graphite lattice, it is decided which graphite is closest as regards the structural state of the free carbon to the sample in hand. By analyzing the lattice parameters (angle Θ_{328}), the composition of the boron carbide is determined [3]. Then the photometric densities of the (002) line of graphite and the (110) of boron carbide are measured by determining the areas of the peaks on the angular-distribution curve of the photometric density. The ratio of the densities S_{002}/S_{110} is found, and this equals the intensity ratio $I_{002\,gr}/I_{110\,b.c.}$, if the work is conducted in the range of photometric densities ≤ 1.

By referring to the calibration curve plotted from the superposition Debye photographs of boron carbide and the carbon-containing material, with due regard to the structural state of the corresponding phases in the sample under analysis, the exposure ratio $t_{gr}/t_{b.c.}$ corresponding to the value of I_{002}/I_{110} so found is determined (Fig. 8). From the ratio $t_{gr}/t_{b.c.}$ on the calculated curve of $C_{free} = f(t_{gr}/t_{b.c.})$ plotted for boron carbide of the same composition as that in the sample under analysis, C_{free} is found (Fig. 9).

The calculated curves of Fig. 9 are plotted [4] in accordance with the system of equations

$$\frac{C_{gr}}{C_{b.c.}} = \frac{(\mu/\rho)_{b.c.}\,t_{gr}}{(\mu/\rho)_{gr}\,t_{b.c.}},$$

$$C_{gr} + C_{b.c.} = 100\%,$$

where C are the weight concentrations of graphite and boron carbide, t is the exposure of the graphite and boron carbide samples on the superposition photographs, and μ/ρ are the mass-

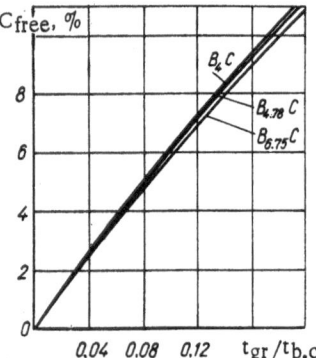

Fig. 9. Calculated curves for determining the amount of free carbon in boron carbide by the superposition method.

absorption coefficients of graphite and boron carbide. In Co K radiation the mass-absorption coefficients for the corresponding components are [1]:

Phase	Graphite	B_4C	$B_{4.78}C$	$B_{6.75}C$
μ/ρ	8.5	5.5	5.4	5.2

If the composition of the boron carbide differs from the compositions B_4C, $B_{4.78}C$, or $B_{6.75}C$, for which the calculated curves have been plotted, the amount of C_{free} is found by extrapolation.

If the series of superposition Debye photographs contains a large number of photographs differing by small degrees of exposure, then without measuring the photometric densities we may visually select one particular x-ray superposition photograph for which the intensity ratio of the lines under analysis is approximately equal to that characterizing the x-ray diffraction pattern under test. For greater accuracy of selection it is recommended that the intensity of the (002) line of graphite should always be estimated by reference to all six of the first lines of boron carbide.

The method has been developed for analyzing samples containing up to 10% free carbon. Quantitative analysis of samples with greater C_{free} contents may be carried out in an analogous manner; however, the (002) line of graphite should then be compared not with the (110) line of boron carbide but with one of the stronger lines: (102), (014), or (201).

The main advantages of the superposition method should be emphasized [4]:

1. There is no need to prepare standard mixtures or introduce additional phases into the sample;
2. The method is independent of chemical analysis (it is only necessary to be confident of the purity and single-phase nature of the components used for obtaining the superposition Debye photographs);
3. The method may be applied to very small quantities of material.

In our own investigations the advantages of the superposition method proved especially valuable in connection with the very small quantities of single-phase boron carbide samples not containing graphite at our disposal, such materials being very rarely achieved in practice.

It was found from the superposition Debye photographs that the lower level of sensitivity for observing free carbon in boron carbide was around 0.15% if its structural state was analogous to Acheson graphite, and 0.3% if the carbon in the boron carbide was extremely finely-dispersed or had a nonequilibrium, poorly-shaped crystal lattice, i.e., if the fine structure of the (002) line of carbon in the sample was the same as in low-graphitized petroleum coke.

The error in parallel determinations based on the superposition Debye photographs is 6% (relative). The reliability of the x-ray analysis is determined by the extent to which the structural state of the free carbon in the sample under analysis agrees with that of the standard materials used for taking the Debye photographs in the superposition series from which the calibration curves are plotted.

The method thus developed may be used for analyzing samples of boron carbide not containing boron nitride in which the carbon phase is finely-dispersed and in an equilibrium state [7], and also in research work associated with the analysis of phase diagrams, in improving the results of chemical analysis, and so on. In particular, the method gives excellent results in the analysis of boron–carbon alloys made by hot pressing.

Literature Cited

1. A. Guinier, X-Ray Diffraction of Crystals [Russian translation], Fizmatgiz, Moscow (1961), p. 592.
2. V. I. Kudryavtsev, Abrazivy, No. 18, p. 152 (1957).
3. V. I. Kudryavtsev and G. V. Sofronov, in: Transactions of a Seminar on Heat-Resistant Materials [in Russian], Vol. 5, Izd. AN Ukr.SSR, Kiev (1960), pp. 52-61.
4. L. S. Palatnik, Izv. Akad. Nauk SSSR, Ser. Fiz., 15(1):134-145 (1951).
5. G. V. Samsonov et al., Boron, Its Compounds and Alloys [in Russian], Izd. AN Ukr.SSR, Kiev (1960), pp. 180-181.
6. G. V. Sofronov and M. I. Sokhor, Certificate Recorded by the State Committee on Inventions and Discoveries, USSR No. 40,332, priority from October 25, 1963.
7. M. I. Sokhor and G. V. Sofronov, in: Transactions of the All-Union Scientific-Research Institute of Analytical Processes [in Russian], Vol. 1, Mashinostroenie, Moscow (1965), pp. 13-18.
8. M. I. Sokhor and V. I. Kudryavtsev, Abrazivy, 4(36):1-9 (1963).
9. M. I. Sokhor, Certificate Recorded by the State Committee on Inventions and Discoveries, USSR No. 40,934, priority from October 25, 1963.

METHOD OF SEPARATING AND DETERMINING THE FREE CARBON IN MATERIALS CONTAINING REFRACTORY COMPOUNDS

L. A. Mashkovich and A. F. Kuteinikov

Scientific-Research Institute of Graphite

The various forms of carbon at present in existence are distinguished from each other by a number of properties, including their relationship to chemical reagents. The so-called amorphous carbon (soot, highly active coals) is easily oxidized by a 30% solution of hydrogen peroxide [10] and adsorbs several dyes (bromothymol blue, etc.) [9], whereas according to our present data graphite remains unaffected by these reagents. In order to conduct a systematic investigation into the chemical properties of various carbon materials and establish their phase relationship, we studied two ways of separating free carbon from refractory compounds: the removal of free carbon by roasting in a muffle furnace at 850°C, and the "wet" oxidation technique.

The first method only proved possible for the SiC–graphite system. In the presence of other carbides, polyfluorethylene resin, or even boron dissolved in silicon, this method was quite unsuitable; in addition to the combustion of free carbon, the carbides were also oxidized, the second process prevailing over the first [6], while polyfluorethylene resin was burnt completely in advance of the carbon at 600-700°C.

In order to choose the conditions for the complete oxidation of various forms of carbon we tried a number of mixtures: 10 ml H_2SO_4 + 10 ml HNO_3 + 5 ml $HClO_4$ + 20 ml $K_2Cr_2O_7$; H_2SO_4 + $K_2Cr_2O_7$ + $HClO_4$ (with various ratios of the components); H_2SO_4 + $HClO_4$; $HClO_4$ + $K_2Cr_2O_7$; H_2SO_4 + $K_2Cr_2O_7$; HIO_3 + H_2SO_4; H_3PO_4 + H_2SO_4 + $HClO_4$ [1, 3, 7, 8, 11].

The oxidizing effect of the mixtures was verified for graphites of various types (PROG-2400, PG-50, AG-1500, etc.), soots (gas channel, thermal, PG-33), and cokes (KNPS, cracking, pitch, aryl resin, etc.). The experiments showed that the forms of carbon studied were only oxidized completely by mixtures Nos. 5 and 2 (ratio of the components 2:2:1). It was found that the presence of perchloric acid (mixture No. 2) worsened the oxidizing power of the mixture, increasing the oxidation time. Subsequently the whole work was carried out with mixture No. 5, comprising a concentrated solution of sulfuric acid and a 5% aqueous solution of potassium bichromate taken in 1:1 ratio.

The mechanism of the oxidation of carbon by the H_2SO_4 + $K_2Cr_2O_7$ mixture is expressed by the equation

$$3C + 2K_2Cr_2O_7 + 8H_2SO_4 \rightarrow 2K_2SO_4 + 2Cr_2(SO_4)_3 + 3CO_2 + 8H_2O.$$

14

TABLE 1. Effect of Boiling Time and Sulfuric Acid Concentration on
the Reduction of Chromium in the Boiling of the Chromic Mixture
(Original Volume 60 ml) without a Reflux Condenser

Volume of chromic mixture after boiling, ml	Ratio of components in mixture: 5% solution of potassium bichromate + 10% sulfuric acid	Boiling time, min	Color of solution	Amount of reduced chromium, %
60	1:1	0	Orange	0.0
51	1:1	60	Orange-red	2.5
40	1:1	120	Dark red	10.0
29	1:1	180	Red-green	30.0
30	1:4	180	Orange	0.8
31	1:2	180	Orange-red	3.0
28	1:1	180	Red-green	20.0
30	2:1	180	Green	80.0
30	4:1	180	Dark green	100.0

In this reaction strict stoichiometry is to be expected (subject to certain conditions), and this may be used to determine the carbon by determining one of the forms of chromium. This approach appears promising in a number of cases, since it enables us to determine carbon over a wide range of concentrations without requiring special apparatus, and offers the possibility of simultaneously determining the carbon and quantitatively separating the insoluble phase of carbide or other inert material.

Our own method of determining carbon is based on this principle [4]. On oxidizing spectrally pure graphite with chromic acid containing various proportions of sulfuric acid and a solution of potassium bichromate, it was found that the time taken for complete oxidation of the graphite diminished with increasing sulfuric acid concentration (Fig. 1). In a dummy experiment (mixture of sulfuric acid and potassium bichromate solution without graphite), redox processes occurred; the color of the solution changed from orange to dark red and then green, indicating the reduction of a certain amount of chromium. The amount of reduced chromium depended not only on the time of boiling and the sulfuric acid concentration but also on the conditions of the dummy experiment (Table 1). On boiling in an open flask, the amount of reduced chromium was higher than on boiling similar solutions with a reflux condenser. Figure 2 illustrates the effect of boiling time and sulfuric acid concentration (sp.gr. 1.84) on the reduction of chromium in the dummy experiment.

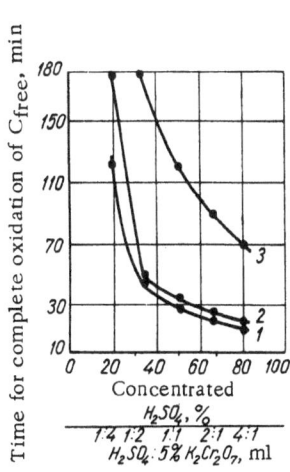

Fig. 1. Time for complete oxidation of the free carbon as a function of sulfuric acid concentration in the chromic mixture: 1) 0.01 g; 2) 0.05 g; 3) 1.0 g.

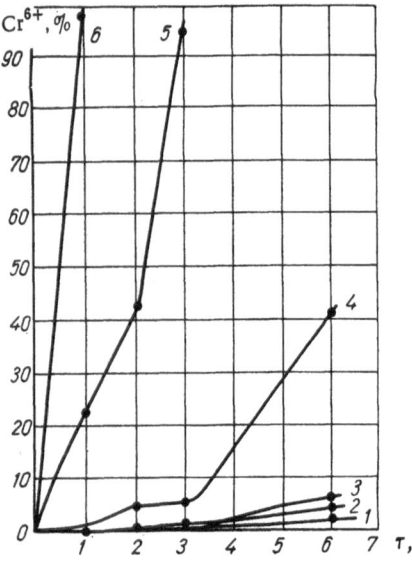

Fig. 2. Effect of boiling time and composition of the chromic mixture on the reduction of chromium in the dummy experiment: 1) $1H_2SO_4 : 4K_2Cr_2O_7$ (10%); 2) $1H_2SO_4 : 2K_2Cr_2O_7$ (10%); 3) $1H_2SO_4 : 1K_2Cr_2O_7$ (5%); 4) $1H_2SO_4 : K_2Cr_2O_7$ (10%); 5) $2H_2SO_4 : K_2Cr_2O_7$ (10%); 6) $4H_2SO_4 : 1K_2Cr_2O_7$ (10%).

In order to prepare the oxidizing mixtures it is best not to employ a 10% solution of potassium bichromate, since a great deal of potassium bichromate is precipitated as these solutions cool. A mixture of sulfuric acid (sp. gr. 1.84) with a 5% solution of potassium bichromate in 1:1 ratio reduces 0.002-0.003 g of chromium in 2-3 h, or 0.2-0.3% of the amount taken.

On prolonged boiling of the solutions (5-11 h) the amount of reduced chromium increases to 0.04-0.07 g, constituting 5-9% of the chromium taken.

A mixture containing 20 vol.% of sulfuric acid oxidizes graphite too slowly and is of no use in practice. The most reasonable composition is a mixture of sulfuric acid and potassium bichromate in 1:1 or 1:2 ratio (the bichromate is used in the form of a 5% aqueous solution). Such mixtures oxidize 1 g of free carbon in 2-3 h. In this period only 0.3% of the chromium is reduced in the dummy experiment. By making a due allowance for the result of the dummy experiment we may eliminate or at least minimize the error arising from this cause.

Subsequently the carbon was oxidized with a 1:1 mixture of sulfuric acid and a 5% solution of potassium bichromate in a flask with a reflux condenser (the dummy experiment was

TABLE 2. Results of a Determination of Free Carbon Based on the Oxidation of the Latter by Bichromate

Amount of spectrally pure graphite, mg	Amount of hexavalent chromium used in oxidizing graphite, mg			Carbon found, mg (arithmetic mean of 10 determinations)	Absolute error, mg	Relative error, %
	Theoretically calculated from the reaction	Found experimentally (arithmetic mean of 10 determinations)	Deviation from calculated value			
5.0	28.9	30.5	+ 1.6	5.3	+ 0.3	5.6
10.0	57.8	58.7	+ 0.9	10.2	+ 0.2	2.0
20.0	115.7	118.1	+ 2.4	20.4	+ 0.4	2.0
40.0	231.1	231.6	+ 0.5	40.1	+ 0.1	0.25
60.0	346.7	344.3	− 2.4	60.0	0	0
80.0	462.2	449.7	− 12.5	77.8	− 2.2	2.75
100.0	577.8	579.0	+ 1.2	100.2	+ 0.2	0.20

TABLE 3. Comparative Data Relating
to the Oxidation of Several
Carbon-Containing Materials

Material studied	Time for complete oxidation, min
PG-33	20-25
Soot	
thermal	20-25
gas channel	20-25
KNPS unroasted	10-20
Coke	
cracking	10-20
pitch	10-20
KNPS graphitized	160-180
Graphite	
PROG-2400	160-180
ZOPG	160-180
PG-50	160-180
ARV	160-180
natural	160-180
AG-1500	160-180

carried out at the same time under the same conditions). A series of experiments on the oxidation of spectrally pure graphite (Table 2) revealed a direct proportionality between the carbon content and the amount of chromium used in its oxidation. Hence the redox reaction between the carbon and the chromium under the conditions envisaged occurs stoichiometrically, and the amount of chromium used in oxidation indicates the carbon content.

Statistical analysis of the results enabled us to determine the relative error of the method, which equalled 4-7% when determining small amounts of carbon (0.04-0.005 g) and 3.5% or better when determining amounts down to 0.1 g.

Analytical Procedure

A 0.2 g sample of material (containing 50% free carbon), crushed so as to pass through a 200-mesh sieve, is treated with 50 ml of a boiling 5% solution of potassium bichromate (exactly measured in a burette) and 50 ml of concentrated sulfuric acid in a conical flask (200-300 ml in volume) with a ground reflux condenser for 2-3 h. A dummy experiment is carried out in parallel. After dissolution, the contents of the flask (with the residue) are transferred to a 250-ml measuring flask.

The residue, constituting refractory compounds, is separated from the solution by centrifuging, washed with water until a neutral reaction is obtained with methyl red, dried in a drying cupboard at 100°C, and weighed.

To aliquot portions (25/250 for the sample, 5/250 for the dummy) of the resultant solution, a 0.05 N solution of Mohr's salt is added from a burette; the excess is titrated with a 0.05 N solution of potassium permanganate [2]. The carbon content is calculated from the formula

$$x = \frac{A \cdot 100}{5.78 \cdot a},$$

TABLE 4

Test substance	Amount of oxidized carbon, %												
	5 min	10 min	15 min	20 min	30 min	45 min	1 h	1 h 15 min	1 h 30 min	1 h 45 min	2 h	2 h 15 min	2 h 30 min
Graphite PROG-2400 (0.02 g)	—	—	—	18.0	—	18.78	19.51	20.62	20.62	17.32	20.46	20.52	21.35
B_4C	3.02	4.2	4.24	3,8	9.2	7.73	8.09	8.84	—	8.73	9.71	—	10.9
Mixture B_4C (0.08g) + +C (0.02g)	14.79	—	17.37	19.57	27.25	23.2	24.67	26.52	25.79	25.91	28.00	28,58	28.8

*Weight of sample 0.1 g.

TABLE 5. Determination of Free
Carbon in Various Materials

Material, g		Free carbon determined	
silicon carbide (boron carbide)	graphite	g	%
—	0.0500	0.0498	99.60
0.5000	—	0.0005	0.10
0.5000	0.0500	0.0502	9.13
1.0000	0.0100	0.0107	1.06
0.1000	—	0.0110	11.0
0.1000	—	0.0106	10.60
0.1000	—	0.0111	11.01
0.5000	—	0.0560	11.20
0.1000	0.0200	0.0305	25.41
0.1000	0.0100	0.0207	18.82
0.1000	0.0300	0.0398	36.15
Polyfluorethylene resin			
0.9000	—	0	0
0.9000	0.1000	0.0986	9.86
0.9200	0.0800	0.0794	7.94

where x is the carbon content in %, a is the sample weight in g, 5.78 is the amount of chromium used in oxidizing 1 g of carbon, A is the amount of chromium used in oxidizing the free carbon in the sample; $A = B_1 - B$. The value of B_1 is determined by titrating the dummy run and B by titrating the sample, using the formula

$$B = [C \cdot v_1 - V] \cdot T \frac{KMnO_4}{Cr} \cdot b,$$

where $C = V'/V''$ is the ratio between the solutions of Mohr's salt (v″) and permanganate (v′), v_1 is the amount of Mohr's salt (in ml) added to the aliquot part, v is the amount of potassium permanganate (in ml) used in the reverse titration, and b is the dilution.

Data relating to the oxidation of various kinds of carbon materials by a mixture of sulfuric

TABLE 6. Comparative Data Regarding the Determination of Free Carbon by
Various Methods

Material under examination	Amount of free carbon according to specification or calculation	Free carbon found		
		by the proposed method	by combustion in a muffle furnace at 850°C	by gas-volumetric method
Silicon carbide	0.12	0.10	—	0.12
Artificial mixture of silicon carbide and graphite	50	50.10	50.0	—
Samples of the graphite silicon carbide system	10	10.13	9.85	9.96
No. 1	—	1.34	1.20	1.27
No. 2	—	61.47	61.10	—
No. 3	—	61.08	60.50	—
No. 4	—	67.02	67.7	—
No. 5	—	81.44	82.1	—
Boron carbide, type II	up to 18	10.60	—	10.07
Sample of the boron–silicon–carbon system	—	63.75	44.82*	62.62
Polyfluorethylene resin[†]	0	0	—	—
Artificial mixture of polyfluorethylene and graphite	10	9.86	—	—
Samples of the polyfluorethylene–graphite system	8	7.94	—	—
No. 1	50	50.30	—	—
No. 2	50	52.10	—	—
No. 3	65	66.80	—	—

* The method based on the combustion of free carbon in a muffle furnace at 850°C cannot be used to analyze samples containing boron, since the oxidation of the boron makes the free-carbon result much too low.

[†] Samples containing polyfluorethylene resin cannot be determined by combustion or gas-volumetrically.

acid and potassium bichromate are presented in Table 3. In view of the fact that for the complete oxidation of structurally free graphite (the most chemically stable of the carbon materials) a period of 3 h was required, the oxidizing characteristics of several refractory compounds were studied under identical conditions (Table 4).

Our experiments showed that some oxides (Al_2O_3, Fe_2O_3, and SiO_2) failed to dissolve in the chromic mixture, while B_2O_3 and P_2O_5 dissolved completely, since in the latter cases the elements had a maximum valence and no chromium was expended in oxidizing them. Silicon, silicon carbide, tantalum carbide, and polyfluorethylene resin neither dissolve nor are oxidized by the mixture, so that their presence in no way interferes with the determination of free carbon by the method described. Boron carbide behaves differently toward the chromic mixture according to the particular method of production and the degree of dispersion. Boron carbide obtained from PROG-2400 graphite crushed to 150-mesh size proved to be fairly oxidation-resistant. The results obtained on determining the amount of free carbon graphically, gravimetrically [7], and by the method here developed were all quite close together.

In view of the fact that the rate of carbon oxidation was much greater than the rate of oxidizing boron carbide, our method of phase separation and determination of the free carbon was completely valid, particularly in the case of large carbon content (over 10%).

The method here developed was tested with artificial mixtures and industrial samples. The results appear in Tables 5 and 6.

Conclusions

1. We have found the optimum conditions for determining carbon by a redox reaction with a mixture of potassium bichromate and sulfuric acid.

2. We have developed a method of determining carbon directly by reference to the redox reaction between carbon and chromium based on the determination of the reduced chromium.

3. The method is suitable for separating and determining the free carbon in materials containing refractory compounds.

Literature Cited

1. G. V. Samsonov (ed.), Analysis of Refractory Compounds [in Russian], Metallurgizdat, Moscow (1962).
2. A. M. Dymov, Technical Analysis of Ores and Metals [in Russian], Metallurgizdat, Moscow (1949), p. 194.
3. Kitakhora, Asakhara, and Atodi, Referat. Zh. Khim. (1959), 78324.
4. L. A. Mashkovich and A. F. Kuteinikov, Soviet Patent No. 184,510, dating from September 10, 1966.
5. L. A. Mashkovich and A. F. Kuteinikov, in: Carbon–Graphite Construction Materials [in Russian], Vol. 2, Metallurgizdat, Moscow (1966).
6. L. A. Mashkovich and A. F. Kuteinikov, in: Transactions of the Conference on Method of Analysis and Testing in the Metallurgical Industry [in Russian], Dnepropetrovsk (1955).
7. G. A. Meerson and G. V. Samsonov, Zavod. Lab., 12:1423 (1950).
8. T. N. Nazarchuk and L. N. Mekhanoshina, Poroshkovaya Met., No. 2, p. 20 (1964).
9. T. N. Nazarchuk and L. R. Pechentkovskaya, Zavod. Lab. 27:256 (1961).
10. N. M. Popova and O. V. Zaslavskaya, Zavod. Lab., 21:1285 (1955).
11. F. Smith, Analyt. Chim. Acta, 13:115 (1955).

STABILITY OF BORON—CARBON COMPOUNDS IN OXYGEN AT HIGH TEMPERATURES

L. E. Pechentkovskaya and T. N. Nazarchuk

Institute of Problems in Materials Science, Academy of Sciences of the Ukrainian SSR

Boron-base alloys are widely used in view of their many useful properties. Nearly always the most important property involved is their resistance to melting.

The high-temperature oxidation of such compounds as carbides and nitrides is complicated by the fact that gaseous oxidation products are formed in addition to solids. The purpose of the present investigation was to study the oxidation of boron—carbon alloys at various temperatures.

According to Moissan [7], oxygen fails to oxidize boron carbide at temperatures up to 500°C, whereas at 1000°C the carbide burns in a flow of oxygen to form CO_2 and B_2O_3.

Ridgeway [8] studied the oxidizing capacity of boron carbide at various temperatures as a function of time. G. A. Meerson and G. V. Samsonov [2] studied the oxidation of boron carbide at various temperatures (820, 1040, and 1080°C). The resultant data regarding the rate of boron carbide oxidation were used as a basis for developing a method of determining the free carbon in this compound.

A study of the oxidation of boron carbide at various temperatures (600-1300°C) [3] showed that the effect began at 700°C. Complete oxidation occurred at 1200-1300°C, the resistance of the boron carbide to oxidation depending on the amount of carbon present.

TABLE 1. Chemical Composition of the Alloys Studied, %

No. of alloy	C_{tot}	B_{tot}	B_{free}
1	13.9	82.7	0.26
2	15.5	82.4	Not found
3	17.9	80.3	" "
4	20.2	79.1	" "
5	22.3	76.8	" "
6	24.8	73.2	0.05
7	27.9	71.3	0.10
8	30.1	68.3	0.10

We studied the oxidizability of alloys belonging to the boron—carbon system in order to determine the effect of chemical composition on their resistance to an oxygen medium. For this purpose we took alloys with varying boron and carbon contents. The samples were analyzed for boric anhydride content as well as free boron and carbon (Table 1). No boric anhydride was found in any of the samples.

The oxidation was carried out in a tubular silit furnace with a constant current of oxygen between 500 and 1300°C, the flow rate being 200 ml/min. The degree of oxidation was judged from the amount of oxidized boron and carbon occurring after a specified period of time.

20

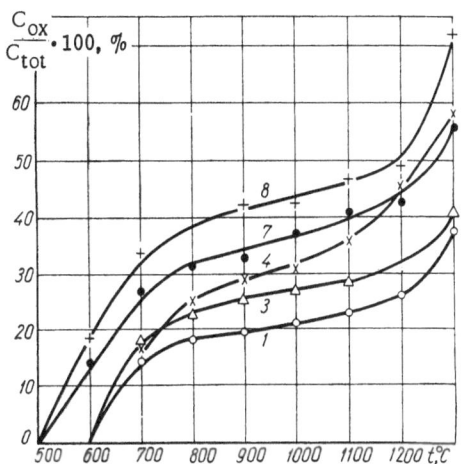

Fig. 1. Oxidation curves of B–C
alloys (referred to carbon). Here
and elsewhere the numbers of the
curves correspond to the numbers
of the alloys in Table 1.

We see from Figs. 1 and 2 that the oxidation of
the B–C alloys starts at 500°C. At this temperature
the oxidation can only be observed by reference to the
formation of B_2O_3. The carbon in the alloy (even for a
concentration of 30.1%) is not oxidized at all at this
temperature. Slight oxidation is only observed at 600°C.
In practice the oxidation of the B–C alloys starts at
700°C. In the range 700-1200°C the boron carbide
undergoes oxidation irrespective of the B:C ratio; this
continues for a specific period and then stops.

A sharp increase in oxidation occurs at 1200-
1300°C, i.e., when the B_2O_3 acquires a substantial vola-
tility. Thus up to a temperature of 1200°C the boron
carbide is to some extent protected from further oxidation by a surface film of B_2O_3.

Although the oxidation curves have a similar
character for all the B–C alloys (Figs. 1 and 2) the
following law is clearly apparent: With increasing
carbon content the resistance of the alloy to oxidation
falls. In our own case the alloy most resistant to oxidation is No. 1 (13.9% C) and the least resistant alloy is No. 8 (30.1% C).

We picture the oxidation of the B–C alloys as follows. At low temperatures (400-600°C)
the boron is oxidized, or at least those of its atoms which occupy the least stable positions in
the lattice. At 600°C the carbon atoms also oxidize. The percentage of oxidized boron at this
temperature is much greater than the percentage of oxidized carbon (Table 2). At 700°C the
carbon undergoes rapid oxidation. The ratio of oxidized boron to oxidized carbon falls, and at
800°C it stabilizes and becomes almost constant (Fig. 3).

We see from Fig. 3 that for the alloys with the more defective structure curves 1, 2, and
3 contain a sharp break. The amount of oxidized boron greatly exceeds the amount of oxidized
carbon. The alloys tend, as it were, to give out the "excess" boron, and the C–B–C line of the
crystal lattice tends to be converted into the C–C–C line of boron carbide $B_{12}C_3$ (= B_4C). For
samples with less defective structure there is no such sharp separation of the boron.

We see from the same figure that the ratio B_{ox}/C_{ox} becomes constant at 800°C. This
leads to the question as to whether all B–C alloys are oxidized up to a certain specific composition, constituting the most stable composition in this
system, or whether each alloy with a specific [B]:[C]
ratio has its own most characteristic structure.

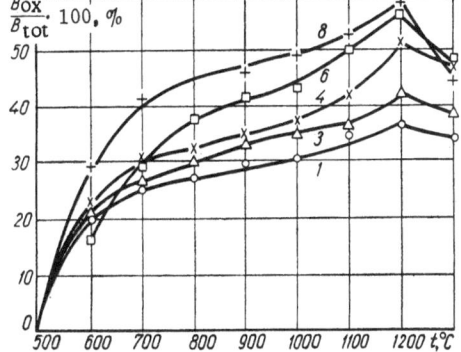

Fig. 2. Oxidation curves of B–C
alloys (referred to boron).

In order to answer this question, B–C alloys
were oxidized with oxygen at 600, 800, 1000, and 1200°C
until oxidation had practically ceased ($\Delta C/\Delta t = 0$). The
samples remaining after the oxidation were washed free
from boric anhydride and dried, and their boron and carbon content was determined. It was considered that, by
this treatment of the alloys, the free boron and carbon
would primarily be oxidized, these constituting the chief
obstacle to the establishment of the composition of
boron–carbon compounds by chemical methods.

TABLE 2. Oxidation of Boron—Carbon

No. of alloy	Amount of C_{tot} in the alloy	Relative percentage							
		B	C	B	C	B	C	B	C
		500° C		600° C		700° C		800° C	
1	13.9	1.4	0	20.0	1.3	25.0	14.2	26.1	18.3
2	15.5	4.5	0	21.3	0.5	26.4	14.6	27.7	15.2
3	17.9	0.8	0	21.5	0.6	28.1	18.5	29.1	22.7
4	20.2	2.4	0	22.0	11.8	30.2	16.2	32.3	25.6
5	22.3	1.5	0	13.3	0.8	29.6	17.3	33.6	19.6
6	24.8	1.5	0	16.3	0.3	30.2	15.3	37.4	31.7
7	27.9	6.0	0	27.8	13.9	33.6	26.9	38.9	31.7
8	30.1	0.4	0	28.6	18.9	42.4	33.2	44.0	37.6

The results of Table 3 show that alloys 1 and 2 (containing 13.9 and 15.5% of carbon in the original samples) tend to form the compound $B_{11}C_2$. On oxidizing the sample with 17.9% of carbon in the original, a compound of composition $B_{4.6}C$ is formed.

On oxidizing alloys with 22.3-27.9% C in the original samples, the oxidation product (after removing the B_2O_3) is a compound corresponding to the formula B_3C for all temperatures and alloys. The alloy containing 30.1% C forms a compound of formula $B_{12}C_5$ ($B_{2.4}C$).

The foregoing experiments convincingly demonstrate that the B–C system is capable of forming four individual substances with compositions corresponding to the formulas $B_{11}C_2$, $B_{4.6}C$, B_3C, and $B_{12}C_5$.

Our own data agree closely with published results. Thus Glasser and Moskowitz [6] considered boron carbide (in view of its similarity to the structure of boron) as constituting solutions of various proportions of carbon in a slightly distorted boron lattice, in which the vacancies offer sufficient accommodation for a maximum of two additional atoms. These authors also explain their own $B_{12}C_5$ compound in this way.

Clark and Hoard [5] suggest that the unit cell of boron carbide offers sufficient room for two additional boron atoms; they consider that the filling of these "holes" with boron atoms leads to the formation of the carbide of composition $B_{14}C_3$, which corresponds to $B_{4.67}C$.

V. I. Kudryavtsev and G. V. Sofronov [1] consider that a continuous series of solid solutions of boron carbide may be obtained thus:

$$B_{12}C_5 \leftarrow B_{12}C_3 \rightarrow B_{14}C_3 \rightarrow B_{15}C_2,$$

where $B_{12}C_3$ has the lattice of boron carbide on which the vacancies ("holes") are empty, i.e., not occupied by boron or carbon atoms. Compounds $B_{12}C_5$ and $B_{14}C_3$ have lattices with two additional carbon atoms and two additional boron atoms lying freely in the vacancies.

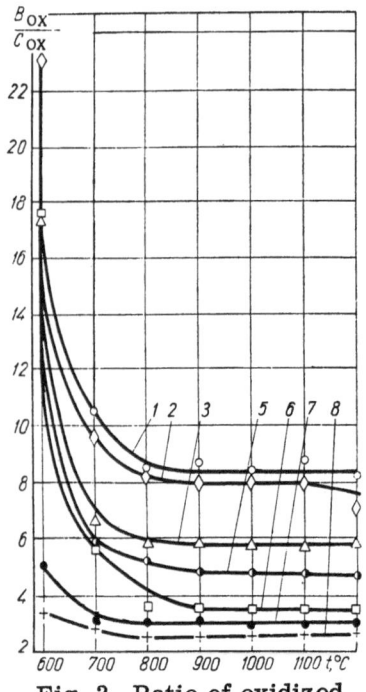

Fig. 3. Ratio of oxidized boron to oxidized carbon at various temperatures.

Thus the existence of compounds of composition $B_{12}C_5$ and $B_{4.6}C$ is confirmed by the foregoing authors. The compound B_3C ($B_{12}C_4$) may be considered as an interstitial solid solution based on the $B_{12}C_3$ cell, in which one additional car-

Alloys at Various Temperatures

of oxidation									
B	C	B	C	B	C	B	C	B	C
900° C		1000° C		1100° C		1200° C		1300° C	
28.7	19.6	30.2	21.5	34.6	23.1	36.0	26.4	34.6	34.7
28.6	19.0	31.0	20.8	32.6	22.0	40.1	31.3	36.2	41.5
34.3	26.0	35.6	27.5	36.6	29.0	42.6	30.5	38.6	44.0
35.0	29.0	36.8	31.1	42.4	36.9	51.5	45.7	47.7	58.4
38.8	27.3	39.4	29.1	43.4	32.9	55.2	44.5	56.1	62.0
40.7	33.6	43.6	36.3	50.6	44.4	57 3	47.7	47.3	55.8
40.8	32.2	41.2	37.9	48.9	42.3	47.0	42.2	46.8	56.3
46.6	42.0	49.0	42.7	52.8	46.7	58.0	48.8	27.3	74.1

TABLE 3. Change in Composition of Boron−Carbon
Alloys on Oxidation at Various Temperatures

Alloy No.	Composition of original sample (before oxidation)		Composition of compound after removing oxidation products				
	B, %	C, %	Computed	600° C	800° C	1000° C	1200° C
1	82.7	13.9	$B_{13}C_2$	$B_{5.6}C$	$B_{5.4}C$	$B_{5.3}C$	—
2	82.4	15.5	B_6C	B_6C	$B_{5.5}C$	$B_{5.5}C$	$B_{5.5}C$
3	80.3	17.9	B_5C	—	$B_{4.6}C$	$B_{4.6}C$	$B_{4.6}C$
4	76.8	22.3	$B_{3.8}C$	$B_{3.9}C$	$B_{3.6}C$	B_3C	$B_{3.1}C$
5	73.2	24.8	$B_{3.2}C$	$B_{3.2}C$	$B_{3.2}C$	$B_{3.2}C$	—
6	71.3	27.9	B_3C	B_3C	B_3C	B_3C	B_3C
7	68.3	30.1	$B_{2.45}C$	$B_{2.5}C$	$B_{2.5}C$	$B_{2.4}C$	$B_{2.4}C$

bon atom has been placed. On the basis of the same set of considerations the compound $B_{11}C_2$ ($B_{5.5}C$) may be considered as a compound with a distorted boron carbide lattice. We remember that Allen [4] mentioned the existence of such a compound with the provisional stoichiometrical formula $B_{5.66}C$.

Conclusions

1. By studying the oxidation of B−C alloys we have established that the resistance of the latter to oxidation diminishes with increasing carbon content.

2. The oxidation of the alloys starts at 500°C. At 800-1200°C the B_{ox}/C_{ox} ratio is constant. For samples with a defective structure there is a sharp oxidation of the boron.

3. Each B−C alloy oxidizes up to a composition characterized by a specific stoichiometric formula; as a result of this the system forms four compounds with the provisional stoichiometric formulae: $B_{11}C_2$, $B_{4.6}C$, B_3C, and $B_{12}C_5$ ($B_{2.4}C$).

Literature Cited

1. V. I. Kudryavtsev and G. V. Sofronov, in: Bulletin of a Seminar on Heat-Resistant Materials [in Russian], Vol. 5, Izd. AN Ukr.SSR (1960), p. 52.
2. G. A. Meerson and G. V. Samsonov, Zavod. Lab., 12:1423 (1950).
3. T. N. Nazarchuk and L. N. Mekhanoshina, Poroshkovaya Met., No. 2, p. 47 (1964).

4.   R. Allen, J. Am. Chem. Soc., 75:3582 (1955).
5.   U. Clark and Hoard, J. Am. Chem. Soc., 65:2115 (1945).
6.   G. Glasser, D. Moskowitz, and B. Post, J. Appl. Phys., 24:734 (1953).
7.   H. Moissan, Compte Rend., 118:556 (1899).
8.   R. Ridgeway, Trans. Electrochem. Soc., 65:117 (1934).

CERTAIN CHEMICAL PROPERTIES OF
BORON CARBONITRIDE

L. E. Pechentkovskaya and T. N. Nazarchuk

Institute of Problems in Materials Science, Academy of Sciences of the Ukrainian SSR

Recently refractory compounds have come to occupy a prominent place in science and technology. One such compound, boron carbonitride, is a refractory material, resistant to the attack of a number of corrosive media; it is comparatively new and unstudied.

In this paper we shall consider a number of chemical properties of boron carbonitride, namely, its interaction with calcium oxide and barium carbonate in a current of oxygen and carbon dioxide, and also its interaction with carbon dioxide itself.

Interaction of Boron Carbonitride with Calcium Oxide and

Barium Carbonate in a Current of Oxygen

We took boron carbonitrides of the following composition:

$$
\begin{array}{llll}
\text{BNC № 1} - 63.5\% - \text{B}; & 12.4\% - \text{C}; & 23.7\% - \text{N} \\
\text{BNC № 2} - 54.4\% - \text{B}; & 5.1\% - \text{C}; & 39.1\% - \text{N} \\
\text{BNC № 3} - 48.3\% - \text{B}; & 5.8\% - \text{C}; & 39.0\% - \text{N}
\end{array}
$$

Finely crushed boron carbonitride powder was mixed in a porcelain boat with calcium oxide previously dried at 120°C (1:4 ratio). The boat was placed in a tubular furnace attached to a source of oxygen, and the cold furnace was blown with oxygen for 1 h. The furnace was switched on, the temperature raised to 900°C, and oxygen was passed for 1.5 h, the gaseous products being collected in a receiver.

At the end of the experiment the gas mixture was analyzed in a VTI-2 analyzer. The cake was dissolved in boiling water. The insoluble residue was filtered off. In the aqueous extract the boron content was determined by titration with alkali in the presence of mannitol by reference to phenolphthalein; the calcium was determined by the oxalic acid method. The oxygen was determined by calculation. Analysis showed that a calcium polyborate of composition CaB_2O_4 passed into the aqueous extract.

In order to determine the composition of the water-insoluble compound, experiments similar to the above were carried out, except that the ratio of the boron carbonitride to the calcium oxide used in sintering was 4:1. The resultant cake was boiled for an hour in water, the insoluble residue was filtered off and treated with boiling 5% HCl; then the calcium and boron contents were determined (Table 1).

25

TABLE 1

№ BNC	Found B^{3+}, g·ion	Found Ca^{2+}, g·ion	O^{2-} content (calc.), g·ion	Calculated formula
1	$2.78 \cdot 10^{-3}$	$0.7 \cdot 10^{-3}$	$4.9 \cdot 10^{-3}$	CaB_4O_7
2	$1.41 \cdot 10^{-3}$	$0.35 \cdot 10^{-3}$	$2.3 \cdot 10^{-3}$	CaB_4O_7
3	$2.76 \cdot 10^{-3}$	$0.7 \cdot 10^{-3}$	$4.9 \cdot 10^{-3}$	CaB_4O_7

Thus, judging by the analysis, a calcium borate CaB_4O_7 poorly soluble in water is also formed. The results of the gas analysis showed that the gaseous reaction products comprised carbon dioxide and nitrogen.

Summarizing the foregoing, the interaction of boron carbonitride and calcium in a current of oxygen may be expressed thus:

$$BNC + CaO + O_2 \rightarrow CaB_4O_7 + CaB_2O_4 + N_2 + CO_2.$$

Analogous experiments were carried out for the reaction associated with the sintering of boron carbonitride and barium carbonate. Some 0.1 g of BNC and 1 g of barium carbonate were mixed in a porcelain boat and sintered for 3 h in a current of oxygen at 800-850°C. The cake was dissolved in boiling water. In the aqueous extract we determined the boron and barium contents (by precipitation as a sulfate). The results appear in Table 2.

Thus on sintering boron carbonitride with barium carbonate in a current of oxygen a polyborate of the composition $Ba_2B_2O_5$ is formed.

In the gas phase the oxygen was accompanied by nitrogen and carbon dioxide. The reaction may thus be written

$$BNC + BaCO_3 + O_2 \rightarrow Ba_2B_2O_5 + N_2 + CO_2.$$

It should be noted that the boron nitride fails to sinter completely with barium carbonate. This is evidently because at the temperature in question a fused bead is formed when barium carbonate interacts with boron carbonitride, the access of oxygen is stopped, and the reaction ceases. Hence in sintering the boron carbonitride it is recommended to use a mixture of barium carbonate and calcium oxide, the latter playing the part of a loosening agent.

TABLE 2

BNC	Found B^{3+}, g·ion	Found Ba^{2+}, g·ion	O^{2-} content (calc.), g·ion	Calculated formula
№ 1	$3.1 \cdot 10^{-3}$	$2.7 \cdot 10^{-3}$	$7.2 \cdot 10^{-3}$	$Ba_2B_2O_5$
№ 1	$5.6 \cdot 10^{-3}$	$5.1 \cdot 10^{-3}$	$13.5 \cdot 10^{-3}$	$Ba_2B_2O_5$
№ 2	$5.8 \cdot 10^{-3}$	$5.5 \cdot 10^{-3}$	$14.2 \cdot 10^{-3}$	$Ba_2B_2O_5$
№ 2	$2.9 \cdot 10^{-3}$	$2.9 \cdot 10^{-3}$	$7.3 \cdot 10^{-3}$	$Ba_2B_2O_5$
№ 3	$1.2 \cdot 10^{-3}$	$1.1 \cdot 10^{-3}$	$2.9 \cdot 10^{-3}$	$Ba_2B_2O_5$

TABLE 3

BNC	Found B^{3+}, g·ion	Found Ca^{2+}, g·ion	O^{2-} content (calc.), g·ion	Calculated formula
№ 1	$2.71 \cdot 10^{-3}$	$0.7 \cdot 10^{-3}$	$4.9 \cdot 10^{-3}$	CaB_4O_7
№ 1	$3.0 \cdot 10^{-3}$	$0.72 \cdot 10^{-3}$	$5.0 \cdot 10^{-3}$	CaB_4O_7
№ 2	$1.42 \cdot 10^{-3}$	$0.35 \cdot 10^{-3}$	$2.5 \cdot 10^{-3}$	CaB_4O_7
№ 2	$2.46 \cdot 10^{-3}$	$0.63 \cdot 10^{-3}$	$4.4 \cdot 10^{-3}$	CaB_4O_7

Interaction of Boron Carbonitride with Calcium Oxide in a Current of Carbon Dioxide

Finely divided boron carbonitride was mixed in a porcelain boat with calcium oxide dried at 120°C (4:1 ratio), placed in a porcelain tube, and connected to a source of CO_2; the latter was passed for 30 min, after which the temperature was raised to 900°C and the CO_2 passed for 2 h, the gaseous products being collected in a receiver. The cake so formed was dissolved in water and boiled for an hour. The insoluble residue was treated with 5% HCl. The amounts of boron and calcium were determined in the aqueous and acid extracts. The results of the acid extract analysis are shown in Table 3.

Thus the interaction of boron carbonitride with calcium oxide in carbon dioxide yields a polyborate of composition CaB_4O_7.

Analysis of the aqueous extract showed that the boron content was in all cases considerably greater than it should be according to the reaction (Table 4); this may have been due to the oxidation of the boron carbonitride by CO_2.

In order to discover the nature of the interaction of boron carbonitride with CO_2 further experiments were carried out. A weighed quantity of boron carbonitride (0.2 g) was held in a flow of CO_2 for 15 min at various temperatures, then treated with boiling water for 1 h, filtered, and the amount of boric anhydride in the filtrate was determined. We see from Fig. 1 that the oxidation of boron carbonitride starts at 700°C and rises sharply on further heating.

Analysis of the gaseous products formed by the boron carbonitride and CO_2 showed that in addition to the CO_2 the mixture contained nitrogen and a great deal of carbon monoxide. The reaction between the carbonitride and the CO_2 evidently took the form:

$$BNC + CO_2 \rightarrow B_2O_3 + CO + N_2.$$

TABLE 4

BNC	Found B^{3+}, g·ion	Found Ca^{2+}, g·ion	Ratio B:Ca
№ 1	$7.3 \cdot 10^{-3}$	$1.2 \cdot 10^{-3}$	6.0
№ 1	$7.1 \cdot 10^{-3}$	$0.7 \cdot 10^{-3}$	9.4
№ 1	$4.0 \cdot 10^{-3}$	$0.8 \cdot 10^{-3}$	5.0

TABLE 5

BNC	Sample weight, g	C_{free} content, g
№ 1 (after reaction with CO_2)	0.2100	$0.51 \cdot 10^{-2}$
№ 2 (after reaction with CO_2)	0.2011	$1.46 \cdot 10^{-2}$
№ 1	0.2006	$0.03 \cdot 10^{-2}$

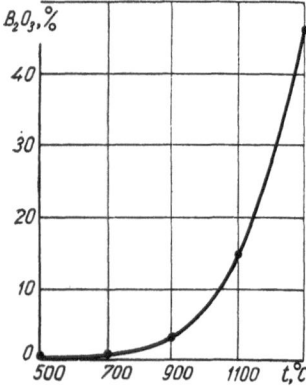

Fig. 1. Oxidation of boron carbonitride in a current of CO_2 at various temperatures.

In addition to this, the cake formed in the boat was covered with a black layer of free carbon, the presence of which was confirmed by combustion of the insoluble residue in a current of oxygen at 600°C. On burning the same amount of boron carbonitride without previous CO_2 treatment, hardly any free carbon was formed (Table 5).

Summarizing the foregoing, the interaction of boron carbonitride with CaO and CO_2 may be written thus:

$$BNC + CaO + CO_2 \rightarrow CaB_4O_7 + C + CO + N_2.$$

Conclusions

As a result of our study of the interaction of boron carbonitride with calcium oxide and barium carbonate in a current of oxygen and carbon dioxide at 800-900°C we have established that:

1) Boron carbonitride interacts with calcium oxide in a current of oxygen in accordance with the equation

$$BNC + CaO + O_2 \rightarrow CaB_4O_7 + CaB_2O_4 + N_2 + CO_2;$$

2) boron carbonitride interacts with barium carbonate in a current of oxygen in accordance with the equation

$$BNC + BaCO_3 + O_2 \rightarrow Ba_2B_2O_5 + N_2 + CO_2;$$

3) boron carbonitride is oxidized in a current of CO_2 at high temperatures;

4) the reaction between boron carbonitride and calcium oxide in a current of CO_2 is governed by the equation

$$BNC + CaO + CO_2 \rightarrow CaB_4O_7 + C + CO + N_2.$$

Literature Cited

1.   G. V. Samsonov et al., Boron, Its Compounds and Alloys [in Russian], Izd. AN Ukr.SSR, Kiev (1960).
2.   M. D. Lyutaya, T. N. Nazarchuk, and K. D. Modylevskaya, Zh. Neorg. Khim., 6:12 (1961).

OXIDATION OF BORON, GALLIUM, AND
INDIUM PHOSPHIDES IN AIR

L. L. Vereikina

Institute of Problems in Materials Science, Academy of Sciences of the Ukrainian SSR

In recent years, the development of methods of obtaining $A^{III}B^{V}$ single crystals by gas-transport reactions [5], in which the original material is often a powdered phosphide, has created interest in the behavior of phosphides in various gas media and their stability toward atmospheric oxygen.

Information regarding the stability of boron, gallium, and indium phosphides is very limited. I. Yu. Andreeva and G. V. Efremov [1] indicate that, on chlorinating and nitriding boron phosphide in the presence of traces of oxygen, BPO_4 is formed. G. A. Goryunova [2] presented some data relating to the initial oxidation temperature of gallium and indium phosphides. Gallium phosphide starts oxidizing at about 875 and indium phosphide at 500°C. However, no quantitative data were presented in regard to the oxidation of the phosphides in either the powdered or compact states.

In order to secure a preliminary estimate of the oxidation of the phosphides, we oxidized boron, gallium, and indium phosphides (0.5 g samples) with oxygen. The oxygen was derived from a cylinder; it passed through a drying system into the reactor. During the heating of the furnace, the boat carrying the sample lay outside the heating zone; on reaching the required temperature it was pushed into the furnace, and the duration of the experiment was reckoned from this instant. The velocity of the oxygen was kept equal to 0.1 liter/min and monitored with a U-shaped liquid flowmeter. Absorption was effected with 60 ml of 0.2 N NaOH + 1 ml of 0.1 N $KMnO_4$ in two or three capillary absorption columns. This arrangement of the absorbing

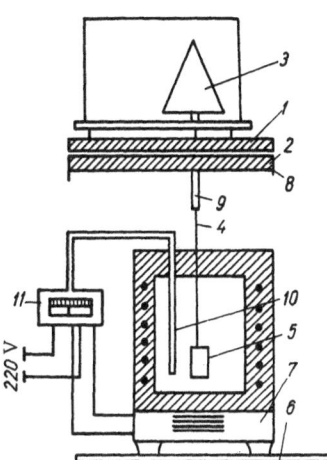

Fig. 1. Arrangement of the apparatus for oxidizing phosphides in air. 1) Marble slab; 2) cantilever; 3) thermobalance; 4) platinum filament; 5) container with sample; 6) slab; 7) vertical muffle furnace; 8) water-cooled slab; 9) guide tube; 10) thermocouple; 11) ÉPD-52 potentiometer.

TABLE 1. Results of the Oxidation of Boron,
Gallium, and Indium Phosphides in Air
($Y = \Delta Q/V$, g/cm^3, where ΔQ = Weight
Increment in g, V = Volume of the Phosphide
in cm^3, τ = Oxidation Time in min)

Phosphide	Temperature, °C	Oxidation equation	Notes
BP	600	$Y^2 = 2.34 \cdot 10^{-6}\,\tau$	Particle
	700	$Y^{2.1} = 9.3 \cdot 10^{-6}\,\tau$	size
	800	$Y^2 = 3.6 \cdot 10^{-6}\,\tau$	2—3 μ
	900	$Y^{2.5} = 6.5 \cdot 10^{-5}\,\tau$	
GaP	700	$Y^{1.7} = 2.7 \cdot 10^{-7}\,\tau$	Particle
	750	$Y^{1.7} = 3.63 \cdot 10^{-7}\,\tau$	size
	800	$Y^2 = 1.32 \cdot 10^{-6}\,\tau$	7—10 μ
	900	$Y^{2.7} = 2.3 \cdot 10^{-6}\,\tau$	Time interval
	950	$Y = 7.1 \cdot 10^{-2}\,\tau$	0—14 min
InP	700	$Y^{2.5} = 1.0 \cdot 10^{-4}\,\tau$	Particle size
	800	$Y^{2.5} = 1.78 \cdot 10^{-5}\,\tau$	7—10 μ
	850	$Y^{2.3} = 5.87 \cdot 10^{-5}\,\tau$	Time interval
	900	$Y = 4.7 \cdot 10^{-2}\,\tau$	0—36 min

TABLE 2. Analysis of the Phosphates and
Activation Energy of Phosphate Formation

Phosphate	Composition, wt.%		Lattice parameters, Å	Activation energy, cal/mole
	P	Me		
BPO$_4$	29.7	10.6	$a = 4.33$; $c = 6.34$	24 500
CaPO$_4$	18.2	43.4	$a = 4.92$; $c = 6.87$	24 750
InPO$_4$	26.3	55.2	—	31 700

columns ensured a uniform distribution of the flow of gas mixture emerging from the system. The solutions of the absorbing columns were analyzed for their phosphorous and metal content. After the oxidation, the residue was analyzed for metal and phosphorus content also; it consisted of metal phosphates. The weight increment largely depended on the surface area of the boat containing the sample; the sample weight and crucible surface area accordingly had to be kept constant.

The absence of gaseous phosphide oxidation products between 600 and 900°C and the formation of phosphates enabled us to use a gravimetric method for studying the oxidation [3-4].

The boron phosphide (0.2 g) and gallium and indium phosphide (0.3 g) samples were placed in a quartz crucible (5 in Fig. 1) with a height of 15 and a diameter of 12 mm, which was suspended in a vertical muffle furnace 7 on a platinum filament 4 situated in a guide tube 9 and connected to a thermobalance 3.

The change in the weight of the sample was determined after a certain period of time without taking it out of the heating zone. Oxidation was continued to constant weight so as to ensure the completion of the oxidation of the sample at the specified temperature. After com-

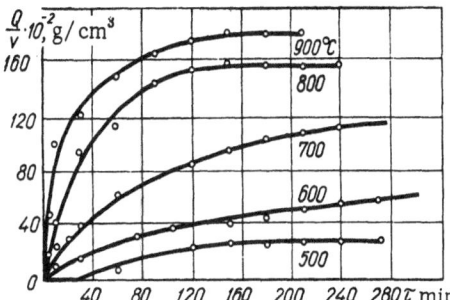

Fig. 2. Oxidation isotherms of
boron phosphide in air.

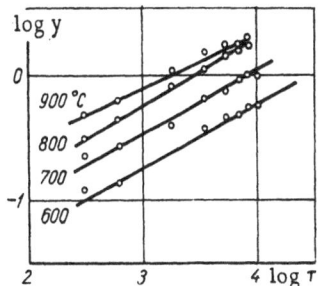

Fig. 3. Results of boron
phosphide oxidation ex-
pressed in log y – log τ
coordinates.

pleting the experiment, the sample was withdrawn from the heating zone, and the reaction products, carefully separated from the residual phosphide, were subjected to chemical and x-ray
analysis. The data relating to the air oxidation of boron, gallium, and indium phosphides are
presented in Tables 1 and 2 and Figs. 2-5.

The results showed that microcrystalline boron phosphide was oxidation-resistant at temperatures up to 600°, gallium phosphide to 700°, and indium phosphide to 700°C. It should be
noted that enlarging the phosphide particles increased th oxidation resistance, so that the three
phosphides constituted valuable semiconductors.

The kinetic oxidation curves of the phosphides obey a parabolic law, $Y^n = K\tau$, in the following temperature ranges: 600-900° for boron phosphide, 700-850° for gallium phosphide, and
700-850° for indium phosphide.

In view of the intense interaction at 900 and 950°C in the case of gallium and indium
phosphides respectively, the process of forming the oxidation products is described by a linear
law in this region. The velocity of the successive stages of the process is controlled by the
diffusion of the reacting components through the layer of phosphates being formed. Chemical
and x-ray analysis showed that the final products were in fact phosphates.

Since our study of the oxidation of BP, GaP, and InP is based on powder samples, the results cannot really yield strictly rational quantitative relationships; the equations obtained are
empirical. This explains the fact that, in the majority of cases, the quadratic (parabolic) oxi-

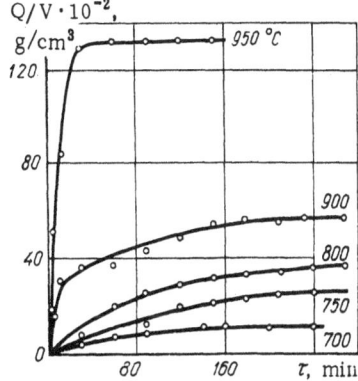

Fig. 4. Oxidation isotherms
of gallium phosphide in air.

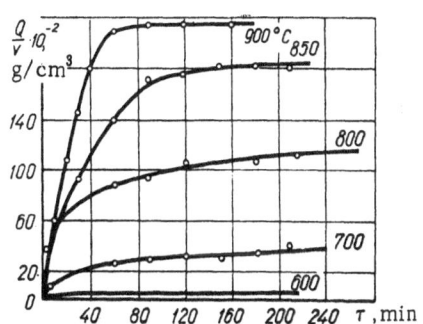

Fig. 5. Oxidation isotherms of
indium phosphide in air.

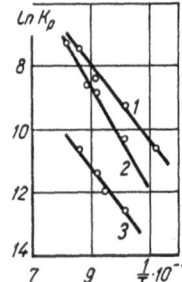

Fig. 6. Velocity constants of the air oxidation of BP (1), GaP (2), and
InP (3) as functions of temperature.

dation law characteristic of compact metal and alloy samples fails to apply as strictly as might
be. However, the results obtained enable us to estimate the onset of oxidation for the phosphides
as well as the final oxidation products. These results may be used in studying the possibility of
obtaining the phosphides in question by indirect methods, and in developing methods of chemical
analysis for the compounds in question and their alloys.

Conclusions

1. Microcrystalline boron phosphide is oxidation-resistant up to 600, gallium phosphide
and indium phosphide to 700°C.

2. The kinetic oxidation curves of the phosphides obey the parabolic law in the following
temperature ranges: BP 600-900, GaP 700-900, InP 700-850°C.

3. Chemical and x-ray analysis shows that the final oxidation products are phosphates.

Literature Cited

1.   I. Yu. Andreeva and G. V. Efremov, Vestnik LGU, Ser. Fiz. i Khim., 10:130 (1964).
2.   G. A. Goryunova, Chemistry of Diamond-Like Semiconductors [in Russian], Izd. LGU,
      Leningrad (1963).
3.   O. Kubashewski and B. Hopkins, Oxidation of Metals and Alloys [Russian translation],
      IL, Moscow (1955).
4.   I. N. Frantsevich, R. F. Voitovich, and V. N. Lavrenko, High-Temperature Oxidation of
      Metals and Alloys [in Russian], GITL Ukr.SSR, Kiev (1963).
5.   H. Schafer, Chemical Gas-Transport Reactions [Russian translation], Mir, Moscow (1964).

HIGH-TEMPERATURE OXIDATION RESISTANCE OF REFRACTORY SILICON NITRIDE—SILICON CARBIDE MATERIALS

I. N. Godovannaya and O. I. Popova

Institute of Problems in Materials Science, Academy of Sciences of the Ukrainian SSR

Silicon nitride and carbide are promising materials for use as refractories; they are highly resistant to mineral acids and alkalis, have a high melting point, and are thermally very stable [1].

We studied the oxidation resistance of refractory materials based on silicon nitride and carbide in various relative concentrations.*

A 0.5 g sample was uniformly distributed in a boat and placed in the porcelain tube of a Mars furnace. The furnace temperature was strictly regulated and monitored with a thermocouple. The velocity of the current of oxygen was constant in all the experiments. The oxygen was passed for a specified period and the amount of oxidized carbon was determined by an absorption-type gas-volumetric method [1]. The content of oxidized carbon was determined from the formula

$$C_{ox} = \frac{(C_{tot} - C_{free}) \cdot 100}{C_{com}}$$

where C_{tot} is the amount of carbon burnt in the time chosen (%), C_{free} is the amount of free carbon in the sample (%), C_{com} is the amount of carbon combined into the form of SiC (%).

In addition to the samples containing both silicon carbide and silicon nitride, the original carbide and nitride were also subjected to oxidation (Table 1).

It follows from the results presented in Figs. 1-6 that at 1000°C all the samples only suffer negligible oxidation (straight line parallel to the horizontal axis). At higher temperatures the degree of oxidation increases with time, although the oxide film protects both the alloys and the original silicon carbide.

If we compare the results with those relating to the oxidation of the original silicon carbide, we find that the oxidiza-

TABLE 1. Chemical Composition of the Samples, Wt.%

Sample No.	SiC	Si₃N₄	C_{free}	C_{com}
1	10.4	89.6	0.11	3.12
2	15.2	84.8	0.21	4.55
3	32.4	67.6	0.11	9.73
4	40.3	59,7	0.13	12.10

*The samples were obtained in the refractory compounds section of the Institute of Problems in Materials Science, Academy of Sciences of the Ukrainian SSR, by V. K. Kazakov.

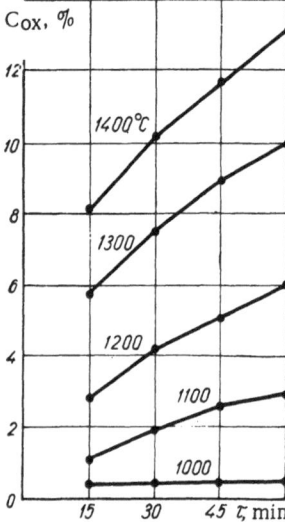

Fig. 1. Oxidation of the original silicon carbide as a function of time over a range of temperatures.

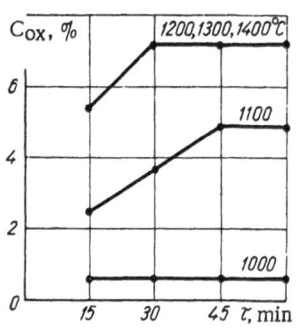

Fig. 2. Degree of oxidation as a function of time for various temperatures of alloy No. 1 containing 10.4% SiC and 89.6% Si_3N_4.

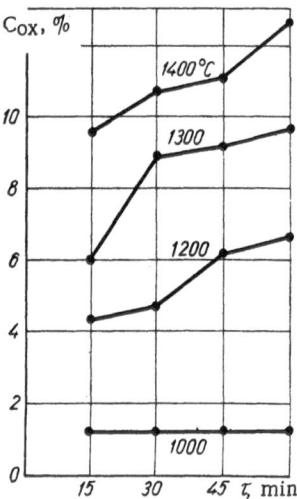

Fig. 3. Degree of oxidation as a function of time for various temperatures of alloy No. 2 containing 15.2% SiC and 84.8% Si_3N_4.

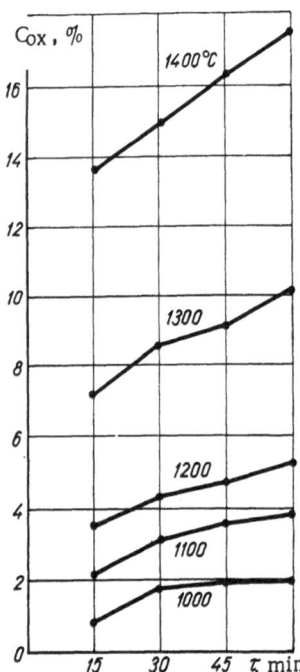

Fig. 4. Degree of oxidation as a function of time for various temperatures of alloy No. 3 containing 32.4% SiC and 67.6% Si_3N_4.

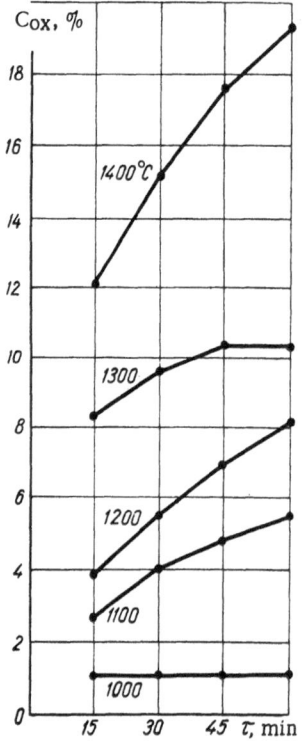

Fig. 5. Degree of oxidation as a function of time for various temperatures of alloy No. 4 containing 40.3% SiC and 59.7% Si_3N_4.

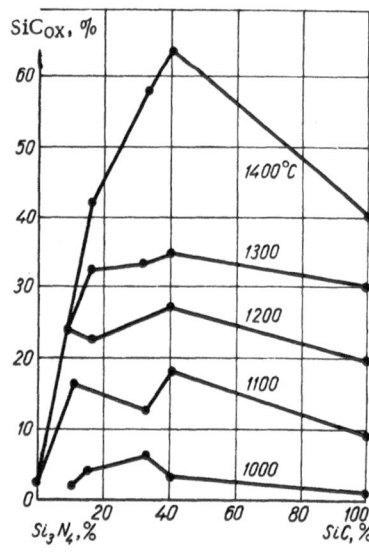

Fig. 6. Oxidation of silicon nitride as a function of composition for various temperatures (holding time 1 h).

bility of the silicon is somewhat greater in the alloys containing the nitride.

The greatest difference in the oxidation of the alloys and the original products occurs at 1400°C for a ratio of $Si_3N_4 \cdot 1.8\ SiC$. Evidently for this composition a compound of minimum oxidation resistance is formed.

Thus the oxidation resistance of refractory alloys comprising silicon carbide and nitride is slightly less than that of the original silicon carbide and nitride.

Literature Cited

1.   Analysis of Refractory Compounds [in Russian], Metallurgizdat, Moscow (1962).
2.   I. S. Kainarskii and É. V. Dekhtyareva, Carborundum Refractories [in Russian], Metallurgizdat, Moscow (1963).

PRODUCTION AND CHEMICAL STABILITY OF THE HYDRIDES OF GROUP IV AND V TRANSITION METALS

M. M. Antonova

Institute of Problems in Materials Science, Academy of Sciences of the Ukrainian SSR

Judged by the character of their chemical bond and physical and chemical properties, the hydrides of Group IV and V transition metals belong to the "metallic" class of materials, although very few data relating to the exact properties of the hydrides are available. The study of these compounds has been retarded by their thermal instability, their peculiar absorption of hydrogen, and the absence of methods of obtaining the hydrides in a compact state. Recently the experimental difficulties in studying the hydrides have been largely overcome, and accordingly a great deal of theoretical and experimental work is now being carried out in connection with the hydrides of transition metals.

The hydrides of the transition metals may be obtained in the form of powders by three quite simple methods. The first of these is the direct interaction of the metals with gaseous hydrogen. For this purpose very pure metals and hydrogen are required. The hydrogen is often obtained by the decomposition of titanium, zirconium, or uranium hydrides [7]. The conditions for producing the hydrides of the transition metals are given in Table 1.

In hydrogenating the compact metal, the product is obtained in the form of fine fragments of hydride, easily broken down into a powder. The specific volume of the hydride is 15-25% greater than the original metal. The hydride is far duller than the metal from which it is derived.

TABLE 1. Conditions for Obtaining Group IV and V Hydrides (2)

Hydride	Hydrogenation temperature, °C	Hydrogenation time, min	Chemical composition of the hydride, wt. %	
			metal	hydrogen
TiH_2	400	30	95.92	4.02
ZrH_2	600	30	97.86	2.14
HfH_2	800	30	98.90	1.10
$VH_{0.9}$	800	120	98.25	1.75
NbH	800	120	98.44	1.06
$TaH_{0.7}$	800	60	99.12	0.38

The second, widely-used method of producing the hydrides of the transition metals is that of reducing the oxides of the refractory metals by metals and hydrides. Calcium hydride (CaH_2) is most frequently used for this purpose [5, 6]. The oxides are loaded in alternate layers with the calcium hydride into an iron tube and then into a furnace, reduction taking place in a hydrogen atmosphere. The temperature of the production process fluctuates between 900 and 1100°C. The resultant hydride is washed free from the calcium oxide mixed with it, by use of a weak HCl solution. In this way titanium and zirconium dihydrides, and vanadium, niobium, and tantalum mono-

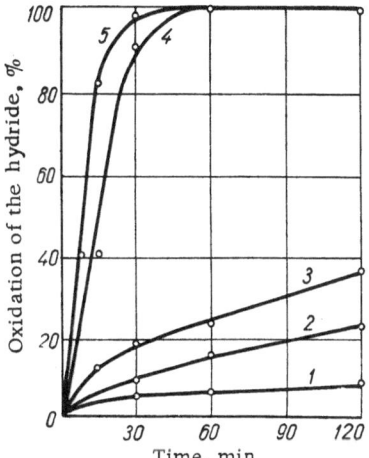

Fig. 1. Kinetic oxidation
curves of titanium hydride.
1) At 400°C; 2) 500°C;
3) 550°C; 4) 600°C; 5) 700°C.

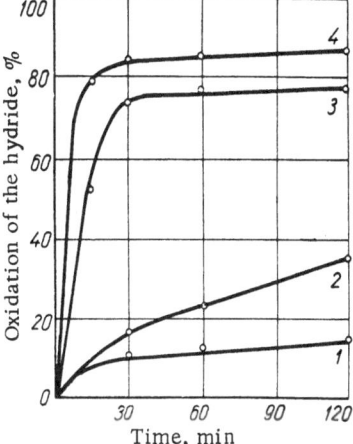

Fig. 2. Kinetic oxidation
curves of zirconium hydride.
1) At 400°C; 2) 450°C;
3) 470°C; 4) 500°C.

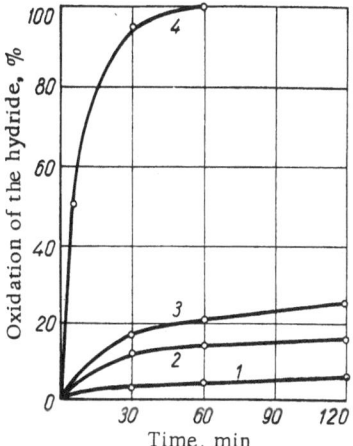

Fig. 3. Kinetic oxidation
curves of niobium hydride.
1) At 300°C; 2) 350°C;
3) 400°C; 4) 500°C.

hydrides are obtained. The first two methods have been quite well developed for producing hydrides on the commercial scale.

The third method of producing hydrides from solutions has not yet been adequately developed. This lies either in treating the metal powder with HF or HCl [9-11] or in making use of the action of phenyl magnesium bromide on the halides of the metals in an ether solution [1]. In the first case a mixture of hydride and metal crystals is precipitated, and it is almost impossible to separate these owing to lack of knowledge regarding the properties of the hydrides. The size of the crystals depends on the concentration of the acid employed. Hydrides obtained by means of phenyl magnesium bromide have also never yet been separated in a state free from ether. At the attempt to remove the ether, the hydrides decomposed into the metal and hydrogen. Thus, if it is to be used on a large scale, the solution method of producing hydrides demands further improvement.

Compact samples of titanium and zirconium hydrides were first obtained very recently [3, 12] by reactive sintering. The method is only suitable for exothermic reactions and is based on the fact that the hydrogenation reaction carried out at the temperature of maximum hydrogen absorption releases a great deal of heat, heating the metal sample being hydrogenated to a temperature of 900-1000°C, as a result of which the ductility of the metal increases, and it sinters by virtue of the heat of the reaction, with simultaneous hydrogenation. The resultant hydride samples have a density of up to 80% of the theoretically calculated value; fracture surfaces appear like the continuous metal, with a sharp metallic luster.

TABLE 2. Composition of the
Hydrides Used for Studying
the Chemical Properties

Hydride	Composition of hydride by chemical analysis, wt.%		Powder size, μ
	metal	hydrogen	
$TiH_{1.53}$	96.7	3.03	15
ZrH_2	97.51	2.00	5—6
NbH	98.70	1.06	10—15

The chemical properties of the hydrides of the transition metals have hardly been studied at all (only a brief qualitative description by Hoard [8]). We therefore decided to start studying the chemical properties of the hydrides, in particular the oxidation of the hydrides in a current of oxygen, and their chemical stability in various widely-used chemical reagents. The data here presented relate to a partial investigation of powdered transition metal hydrides. The hydrides were obtained by the direct interaction of the metal powders with hydrogen. The compositions of the hydrides are indicated in Table 2.

TABLE 3. Stability of Zirconium Hydride
in Chemical Solutions

Medium	Temperature, °C	Residue, %	Dissolved substance, %	Metal in solution, %
H₂O	100	100	—	—
HCl conc.	108	96.4	3.60	3.60
HCl (1:1)	108	95.4	4.60	4.50
HNO₃ conc.	110	97.8	2.20	2.25
HNO₃ (1:1)	110	97.8	2.20	2.20
H₂SO₄ conc.	280	—	100	100
H₂SO₄ (1:1)	136	68.95	31.05	30.95
HF conc.	25	—	100	100
H₃PO₄ conc.	100	—	100	100
H₂O₂ (33%)	100	99.35	0.65	0.65
NH₄OH	25	100	—	0.37
NaOH (10%)	110	102.2	—	30.7
KOH (10%)	105	100.8	—	3.05
Acetic acid	100	100	—	1.41
Tartaric acid	25	100	—	—
Oxalic acid	100	93.72	6.28	6.27
CCl₄	25	100	—	—
Acetone	25	100	—	—
Ethyl alcohol	25	100	—	—
HF + HNO₃ (1:1)	25	—	100	100
HNO₃ + H₂SO₄ (1:1)	100	85.8	14.2	14.15
HNO₃ + H₂O₂ (1:1)	100	97.7	2.3	3.10

The rate of oxidation of titanium, zirconium, and niobium hydrides was studied by burning hydride samples of 1 g in weight in a quartz reactor placed in a resistance furnace in a current of oxygen; the water vapor formed as a result of oxidation was determined gravimetrically, and the hydride was weighed before and after the experiment. The degree of oxidation of the hydride was estimated both from the amount of water vapor trapped and from the weight increment of the oxidized hydride. The hydride powders were oxidized for 2 h at 300-700°C. In order to obtain comparable results from the combustion of all three hydrides, the amount of oxidized hydrogen and the total initial hydrogen content were taken as 100% in each hydride for purposes of calculation. The results of the oxidation are shown in Figs. 1-3. The form of the resultant oxidation curves gave the law of oxidation, and the temperature dependence of the oxidation constant was also calculated from this.

The results of the experiments showed that the oxidation of titanium hydride started from 500-550°C, that of zirconium hydride from 450-470°C, and that of niobium hydride from 300-350°C. Analysis of the kinetic oxidation curves of the hydrides shows that the law of oxidation changes as the oxide film grows. The formation of each oxide is reflected in its own oxidation law. As the oxide is converted into a higher form, the oxidation law acquires the form of a quadratic parabola, which indicates that the oxidation process is controlled by the diffusion of oxygen through the film of higher oxide. The kinetic oxidation curves of the hydrides calculated from the weight of trapped water vapor and from the weight increment of the hydride differ qualitatively and quantitatively. We may suppose that the oxidation of the hydride is preceded by its decomposition, with subsequent separate oxidation of the metal and the released hydrogen. This is indirectly supported by data relating to the dissociation of the hydrides in question in vacuo [4].

The stability of the hydrides in various media was studied at room temperature and boiling point. In studying the stability of the hydrides at room temperature, 0.2 g of hydride was poured into a 100-ml beaker of the reagent and held for 8 h with periodic agitation. In study-

TABLE 4. Stability of Titanium Hydride
in Chemical Solutions

Medium	Temperature, °C	Residue, %	Dissolved substance, %	Metal in solution, %
H_2O	100	100	—	—
HCl .conc.	100	—	100	100
HNO_3 conc.	25	59.8	40.2	40.2
H_2SO_4 conc.	280	—	100	100
H_2SO_4 (1:1)	120	—	100	100
HF conc.	25	—	100	100
H_3PO_4 conc.	70	—	100	100
H_3PO_4 (1:1)	100	3.97	96.03	88.70
NH_4OH	100	108.5	—	0.44
KOH (10%)	100	112.2	—	—
NaOH (10%)	100	111.8	—	43.8
Tartaric acid	100	100	—	—
Acetic acid	100	100	—	—
Oxalic acid	100	5.58	96.42	97.2
CCl_4	25	100	—	—
Acetone	25	100	—	—
Dichloroethane	25	100	—	—
HF + HNO_3 (1:1)	25	—	100	100
H_2SO_4 + HNO_3 (1:1)	100	—	100	100

ing the stability at boiling point, the sample was placed in a conical flask connected by means of a ground-glass joint to a reflux condenser and boiled for 8 h. The insoluble residue was filtered in a Schott No. 4 filter and the percentage of undissolved metal was determined from the weight loss. The metal passing into solution was determined by the Cupferron method (Table 3).

The zirconium hydride was completely decomposed by concentrated sulfuric and hydrofluoric acids of all concentrations with the vigorous evolution of small hydrogen bubbles. On interacting with orthophosphoric acid the reaction proceeded more quietly and only took place at boiling point, no hydrogen evolution being visually observed. The remaining mineral acids dissolved zirconium hydride weakly. In the dissolution of the hydride in acid solutions, the zirconium salts of the corresponding acids were formed. No hydrolysis of the salts was observed on diluting the acids. Mixtures of acids failed to increase the solubility of zirconium hydride even at boiling point. Only in HF mixtures did the ZrH dissolve completely.

On dissolving zirconium hydride in alkalis there is usually an increment in the weight of the residue as compared with the original sample, and considerable amounts of zirconium appear in the solution. This may be explained by the formation of zirconates, the solubility of these depending very greatly on the acidity of the medium, so that some of the zirconium passes into solution, while the weight of the residue increases as a result of the formation of the insoluble salt. The external appearance of the powder remains unaltered on interaction with alkalis.

In organic acids and solvents zirconium hydride dissolves poorly, a marked transfer of zirconium into solution only occurring for oxalic acid (Table 4).

Titanium hydride easily dissolves in hydrochloric, sulfuric, and hydrofluoric acids of all concentrations, even in the cold. On boiling the dissolution accelerates sharply. Titanium hydride is also completely dissolved in orthophosphoric acid. Dissolution in concentrated sulfuric acid occurs in a very special manner: First there is a vigorous reaction with the evolution of gas, and a white residue in the form of a very fine powder is formed; on further boiling in sulfuric acid this dissolves completely with the formation of a colorless transparent solution.

TABLE 5. Stability of Niobium Hydride
in Chemical Solutions

Medium	Temperature, °C	Residue, %	Dissolved substance,%	Metal in solution, %
H_2O	100	100	—	—
HCl conc.	108	61.1	38.9	38.8
HCl (1:1)	108	96.8	3.2	2.96
HNO_3 conc.	110	99.4	0.6	0.5
HNO_3 (1:1)	100	98.8	1.2	1.32
H_2SO_4 conc.	280	—	100	100
H_2SO_4 (1:1)	136	43.35	56.65	57.1
H_3PO_4 conc.	100	5.39	94.61	94.70
H_3PO_4 (1:1)	100	69.2	30.8	30.8
HF conc.	25	—	100	100
KOH (10%)	100	27.8	72.2	85.0
NaOH (10%)	100	197.5	—	40.2
H_2O_2 (33%)	100	92.9	7.1	7.15
Tartaric acid	100	96.8	3.2	3.25
Acetic acid	100	100	—	—
Oxalic acid	100	60.6	39.4	39.5
CCl_4	25	100	—	—
Acetone	25	100	—	—
Dichloroethane	25	100	—	—

A similar kind of dissolution occurs in a mixture of sulfuric and nitric acids. We may suppose that in this case TiO_2 or $Ti(SO_4)_2$ is formed, and this then dissolves with the formation of titanyl sulfate $TiOSO_4$, which exists in acid solutions. In order to elucidate the composition of the white deposit we attempted to separate it from the main solution, but this proved impossible. After separation of the deposit on the Schott filter and washing with hot distilled water, the residue completely dissolved in the latter. This excluded TiO_2 as a possible identification and indicated that on dissolving titanium hydride in concentrated sulfuric acid a partly-soluble $Ti(SO_4)_2$ was first formed and that on further boiling this was converted into the easily-soluble titanyl sulfate $TiOSO_4$.

In sulfuric and orthophosphoric acids a colorless solution is formed, suggesting the tetravalent ion Ti^{4+}. In all the other inorganic acids and mixtures of these, the ion of tervalent titanium is formed (violet solution). Of the organic acids and solvents studied, only oxalic acid has a marked effect on the passage of titanium into solution in appreciable quantities.

The manner in which titanium hydride dissolves in alkalis is the same as in the case of zirconium hydride. In this case titanates of fairly low solubility are formed. This are precipitated and yield a mixture with the original hydride powder.

Niobium hydride exhibits a lower solubility in acids than titanium and zirconium hydrides. Even in concentrated sulfuric acid it requires 1.5 h to dissolve at boiling point. This indicates that niobium has a more amphoteric character in solution, exhibiting a greater tendency to form salts with alkalis. In caustic soda solutions the hydride is dissolved completely; however, on diluting the solution the resultant salt is hydrolyzed with the formation of a fine-crystalline white residue of hydroxide, which then only dissolves in acids and alkalis with great difficulty. The relation of niobium hydride toward organic acids is exactly the same as that of titanium and zirconium hydrides.

Following our investigations into the stability of titanium, zirconium, and niobium hydrides in chemical reagents, it may readily be seen that the resistance of the hydrides is exactly the same as that of the original metals; this can nevertheless only be asserted qualitatively, as there is no evidence relating to the stability of metal powder under the same conditions.

The investigation has shown in particular that the dissolution of the hydrides in chemical reagents cannot be simply reduced to the idea of "decomposition of the hydride into metal and hydrogen followed by dissolution of the metal in the solution." Analysis of the insoluble residue for metal and hydrogen shows that the composition of the hydride never remains constant during the interaction; it changes, yet in no case is a complete absence of hydrogen apparent. The amount of hydrogen remaining in the residue is different for the dissolution of the hydride in different reagents; however, no regular laws governing its quantitative proportions are to be seen.

Literature Cited

1. O. I. Alekseev and E. I. Krylov, Ukrain. Khim. Zh., 22:2 (1956).
2. M. M. Antonova and G. V. Samsonov, Zh. Prikladnoi Khim., 38(11):2393 (1965).
3. M. K. Antonova, Poroshkovaya Met., No. 7, p. 25 (1965).
4. M. M. Antonova, Ukrain. Khim. Zh., 32:6 (1966).
5. G. A. Meerson, G. A. Kats, and A. V. Khokhlova, Zh. Neorg. Khim., 13:1771 (1940).
6. G. A. Meerson and O. P. Kolchin, in: Transactions of the Ministry of Nonferrous Metals and Gold [in Russian], No. 25, Metallurgizdat, Moscow (1955).
7. V. I. Mikheeva, Hydrides of the Transition Metals [in Russian], Izd. AN SSSR, Moscow (1960).
8. G. Hoard, Introduction to the Chemistry of Hydrides [Russian translation], IL, Moscow (1955).
9. G. Brauer and H. Müller, J. Inorg. Nucl. Chem., Vol. 17, No. 1/2 (1961).
10. W. Espe, Zirconium Fussen (Bauern), 9:11 (1953).
11. A. Jamagut and M. Otsuke, Z. anorg. allg. Chem., 29:131 (1957).
12. Z. Motr, Ber. Dtsch. Keram. Ges., 40:91 (1963).

CHEMICAL ANALYSIS OF THE REACTION PRODUCTS OF BORON WITH ARSENIC AND PHOSPHORUS

A. A. Reshchikova and Z. S. Medvedeva

N. S. Kurnakov Institute of General and Inorganic Chemistry, Academy of Sciences of the USSR

Boron arsenides and phosphides are refractory compounds offering certain uses in semi-conductor technology. The synthesis of these compounds has been a matter of great concern in the Soviet Union and elsewhere over the last few years [10]; however, very little has appeared in the literature regarding the chemical analysis of these substances [1, 9, 12].

A study of the processes taking place in the interaction of boron with phosphorus and arsenic led to the development of methods of synthesizing powdered boron phosphide BP [3], boron arsenide BAs, and arsenic hexaboride B_6As [6, 4], as well as solid solutions BAs–BP of the ternary B–As–P system [5]. The products so formed were determined by chemical, spectral, and x-ray phase analysis.

In this paper we shall present our own methods of analyzing the products of the interaction of boron with arsenic and phosphorus and also ternary powders of the B–As–P system. The original substances for the synthesis were amorphous boron, red phosphorus, and crystalline arsenic (Table 1). The interaction of solid boron with phosphorus and arsenic vapor was conducted under heterogeneous conditions.

The interaction of amorphous boron with arsenic was studied in a vertical furnace in evacuated and sealed quartz ampoules, the original composition of the mixtures being widely varied. The temperature in the furnace was 700-760 and 1100°C, depending on the composition of the mixture. The results of x-ray [4] and chemical [7] analysis of the reaction products showed that boron and arsenic formed two chemical compounds: BAs and B_6As. In the presence of more than the stoichiometric proportion of arsenic (over 50 at.%) and at a temperature of 700-760°C, a black powder (BAs) was formed; this was easily separated from the arsenic by evaporating the latter at 400°C into the "cold" part of the ampoule. At 1100°C boron reacted

TABLE 1. Spectral Analysis of the Original Materials

Substance analyzed	Impurity, wt.%					
	Si	Te	Mg	Pb	Al	Cu
Boron	$3 \cdot 10^{-3}$	$1.3 \cdot 10^{-3}$	$3 \cdot 10^{-4}$	$5 \cdot 10^{-4}$	$2.7 \cdot 10^{-4}$	$4.5 \cdot 10^{-4}$
Phosphorus	$3.6 \cdot 10^{-4}$	$1.8 \cdot 10^{-4}$	—	—	$1 \cdot 10^{-4}$	$1 \cdot 10^{-4}$
Arsenic	10^{-5}	10^{-5}	—	—	10^{-5}	—

TABLE 2. Chemical Stability of B$_6$As
(wt.%) in Certain Acids
(Sample 100 mg, Solvent 20 ml)

Medium	On boiling for 2.5 h	In the cold for 1 day
HNO$_3$ ($d=1.4$ g/cm^3)	28.7	35.0
HNO$_3$ + H$_2$O$_2$ (1:1)	43.0	36.1
HNO$_3$ + H$_2$O$_2$ ($d=1.19$ g/cm^3)	2.9	2.5
HCl + HNO$_3$ (3:1)	41.1	48.0
H$_2$SO$_4$ ($d=1.8$ g/cm^3)	8.5	5.0

with a certain quantity of arsenic (taken so as to be less than 50 at.%) and formed a yellow-brown powder (B$_6$As), either as a single phase or as a system containing an excess of boron or arsenic, depending on the composition of the original mixture. The hexaboride was separated from excess boron by boiling in nitric acid, in which the B$_6$As was almost insoluble, and from excess of arsenic by distillation without opening the ampoule, as in the case of BAs.

The pure hexaboride was also obtained from boron arsenide by roasting the latter at 1100°C. As a result of thermal dissociation the BAs loses some of its arsenic and transforms into B$_6$As. The arsenic thus separating is distilled into the "cold" part of the ampoule.

The resultant boron–arsenic compounds are characterized by great chemical stability. Verification of the solubility of B$_6$As in several acids (Table 2) showed that none of the acids or acid mixtures taken dissolved the hexaboride completely.

On prolonged boiling, boron arsenide is almost completely soluble in nitrose; however, the presence of an insoluble residue and the long period of boiling (8-10 h) make this method undesirable for analytical purposes.

In order to achieve a dissolved state the method of fusion was employed. For this purpose 0.1 g of sample powder was fused in a nickel crucible with 1 g of alkali and 0.5 g of sodium peroxide. The alkali was first melted to remove moisture; then it was cooled and again melted after introducing the sample, thus reducing the activity of the latter. After solidification, sodium hydroxide was introduced into the crucible and carefully melted, in a weak burner flame, starting by heating the sides of the crucible. As the mixture melted, the flame was made larger and heating continued for 10 min. After leaching the melt with hot water, the boron and arsenic were brought into solution in the form of a borate and arsenate. In order to decompose the peroxide, the solution was boiled for 1 h in a flask with a reflux condenser, then filtered and transferred to a 200 ml measuring flask.

The arsenic was determined iodometrically. The presence of boron interfered in no way with the determination. To an aliquot portion 50 ml in volume we added 10 ml of H$_2$SO$_4$ (1:1) and several drops of a permanganate solution until a rose color appeared. The excess of permanganate was decomposed by introducing 2 ml of alcohol and boiling. After the decomposition of the remains of the peroxide in this manner, 40 ml of HCl were added ($d=1.19$ g/cm^3) together with 0.5-1 g of KI. The iodine so formed was immediately titrated with a 0.02 N solution of thiosulfate until the solution was decolored. The titration was completed without using starch, as the latter gave unreliable colorings in the case of strong acidity. After 20 min the solution was further titrated, if on standing in darkness it again acquired a yellow hue. A dummy experiment was carried out with the same amounts of reagents.

The boron was determined from another portion of solution by titrating with alkali in the presence of mannitol. The end point was noted visually by reference to phenolphthalein, or else potentiometrically. For an amount of boron in the sample corresponding to the composition BAs (B 12.72, As 87.28 wt.%) the titration was carried out thus. To an aliquot portion 50 ml in volume we added several drops of phenolphthalein after adding HCl so as to achieve a rose tint with methyl red, boiling for 5-7 min to remove carbonic acid, and cooling rapidly in running water; we then established the neutral point by adding a 0.1 N solution of NaOH. After introducing 1-2 g of mannitol we titrated with alkali until a stable rose tint appeared, not vanishing on adding more mannitol.

TABLE 3. Determination of Boron in the Presence of Arsenic from an Artificial Mixture of Boric and Arsenic Acids

Boron, mg	Added arsenic, mg	Calculated correction, mg	Boron determined, mg	Error of determination mg	Error of determination %
8.0	—	—	8.0	0.0	0.0
8.0	—	—	8.0	0.0	0.0
8.0	—	—	8.0	0.0	0.0
8.0	9.8	1.20	8.1	+0.1	1.2
8.0	9.8	1.20	8.0	0.0	0.0
8.0	9.8	1.20	8.1	+0.1	1.2
16.0	—	—	16.0	0.0	0.0
16.0	—	—	15.9	−0.1	0.6
16.0	24.5	2.29	15.9	−0.1	0.6
16.0	24.5	2.99	16.1	+0.1	0.6

If the boron content corresponds to the composition of arsenic hexaboride (B 46.43, As 53.57 wt.%), the foregoing method of titration gives too low a value for boron owing to the partial neutralization of the boric acid when establishing the neutral point by reference to phenolphthalein. The analysis of arsenic hexaboride is therefore conducted slightly differently.

The neutral point was established by reference to methyl red, this corresponding to the neutralization of the strong hydrochloric acid and the first stage of the weak arsenic acid. Then mannitol was introduced together with phenolphthalein, and the solution was titrated with alkali until it acquired a rose color.

Since on titrating with respect to phenolphthalein the arsenic acid was titrated in the second stage at the same time as the boric acid, a correction for arsenic was introduced when calculating the boron content; this was based on the iodometric determination of the arsenic carried out on the other portion of the solution. The amount of thiosulfate used in the titration was converted to the corresponding volume of alkali. The latter was subtracted from the total volume of alkali used in determining the boron.

The results of a determination of boron from an artificial mixture with arsenic are presented in Table 3. In determining the boron potentiometrically the fixed pH method was employed, this eliminating the interference arising from the presence of the weak acids [1, 2]. For this purpose the pH of the solution was set at 6.9 on the LP-58 potentiometer after boiling and cooling. Then the mannitol was introduced and the solution was titrated again with 0.1 N alkali to pH 6.9. The titer of the alkali was established by reference to twice-recrystallized boric acid under the same conditions. This method was verified for artificial mixtures of boric acid with arsenic acid, and also with a mixture of phosphoric and arsenic acids (Table 4). The error in determining the boron was no greater than ± 0.4 rel.%.

In calculating the combined boron and arsenic in hexaboride powders, the content of free elements which had not reacted was taken into account; these were determined after boiling the sample in hydrogen peroxide with the addition of several drops of nitric acid. Gueilleron and Thevenot [12] determined the B:As ratio in this compound as 6.2-6.3 by chemical analysis; this failed to give a clear choice between the formulas B_6As and $B_{13}As_2$.

On analyzing several samples of arsenic hexaboride powder obtained in our own laboratory, we found a B:As ratio of nearly 6:1, and accordingly decided on the formula B_6As. This composition is supported by x-ray analysis [4].

The composition of the black powder obtained at 700-760°C was almost stoichiometrical BAs according to chemical analysis. In studying the reaction between boron and phosphorus we analyzed the kinetics of the process. The experiments were carried out in a two-zone horizontal furnace in which the constant temperature of the phosphorus zone was 400°C and the variable temperature of the boron zone 1000-1150°C. The boron and phosphorus (in stoichiometric ratio) were placed in quartz boats and set at opposite ends of the ampoule, which was evacuated to 10^{-5} mm Hg and sealed. After an appropriate holding period, the ampoule was extracted from the furnace and the phosphorus which had not entered into the reaction was condensed in the "cold" part at 800°C. After opening the ampoule, the boat which had contained the boron was weighed, and the yield of the reaction was calculated from the increment. Since the reaction never proceeded to completion, the resultant powder was subjected to chemical analysis. Ex-

TABLE 4. Potentiometric Determination of Boron
in the Presence of Arsenic and Phosphorus

Taken, mg			Boron determined, mg	Relative error, %	Boron taken, %	Boron determined, %	Absolute error, %
boron	phosphorus	arsenic					
27.1	—	—	27.11	+ 0.04	100	100.04	+ 0.04
27.1	—	—	27.13	+ 0.11	100	100.11	+ 0.11
27.1	—	—	27.05	— 0.19	100	99.82	— 0.19
27.1	36.4	—	27.45	+ 1.30	42.68	43.23	+ 0.55
27.1	36.4	—	27.09	— 0.04	42.68	42.67	— 0.01
27.1	36 4	—	27.13	+ 0.11	42.68	42.73	+ 0.05
27.1	—	37.3	27.06	— 0.15	42.08	42.02	— 0.06
27.1	—	37.3	27.10	0.0	42.08	42.08	0.0
27.1	36.4	37.3	27.22	+ 0.44	26.88	27.00	+ 0.12
27.1	36.4	37.3	27.30	+ 0.74	26.88	27.08	+ 0.20
13.55	29.85	29.78	13.47	— 0.60	18.69	18.58	— 0.11
13.55	29.85	29.78	13.50	— 0.40	18.69	18.62	— 0.07
13.55	29.85	29.78	13.64	+ 0.60	18.69	18.82	+ 0.13
13.55	29.85	29.78	13.49	— 0.40	18.69	18.61	— 0.08

Note. Mean error of the determination, %: relative ± 0.36, absolute ± 0.12.

periments showed that the yield of BP depended not only on the time and temperature of synthesis but also on the surface area involved in the interaction. The boron phosphide obtained in the form of a crystalline, light-brown powder was separated from the boron which had not reacted by boiling in a concentrated mixture of HNO_3 with H_2O_2, or in aqua regia, in which boron phosphide was insoluble.

The chemical stability of boron phosphide has been noted by many authors [1, 10, 13]. In order to decompose this compound Andreeva and Efremov [1] tried chlorination in an isolated system at 550°C, with additional melting of the nonreacting material in soda and saltpeter. Sentyurina's method [9] of bringing boron phosphide into solution by boiling with a mixture of HNO_3 and $NaNO_3$ also provides for the subsequent alkali fusion of the undissolved residue. However, the most convenient procedure would appear to be a method in which the whole sample is decomposed without additional fusion. Thus on melting with a mixture of alkali and sodium peroxide, or with just the peroxide, the products so formed are completely soluble in water. However, in both cases losses are inevitable owing to the formation of phosphine [13]. On melting boron phosphide with a mixture of soda, potash, and nitrate, no ignition occurs if a reasonable excess of flux is employed. This method was used by Samsonov and Titkov [8], who took a 80-100-fold excess. Tests showed that on reducing the excess of flux by a factor of two, complete decomposition of the sample, with dissolution in hot water, was also achieved.

In order to determine the phosphorus, we employed a version of indirect complexonometric titration after deposition of the phosphate ion with a titrated solution of bismuth nitrate at a concentration of 4 ml HNO_3 (d = 1.4 g/cm³) in 100 ml of solution. The precipitation and coagulation occupied no more than 20-25 min. Then the deposit of bismuth phosphate was separated, dissolved in hot dilute HNO_3 (1:1), and after establishing a pH of 1.0-1.1 by means of NH_3 (1:1) the bismuth was titrated with Complexone III, using xylenol orange as indicator. The precipitation of the phosphate ion with bismuth nitrate may be carried out after the potentiometric determination of boron by titration with alkali and mannitol.

The results of the determination of phosphorus from an artificial mixture of phosphate with boric acid are shown in Table 5. The composition of the resultant boron phosphide (based on the chemical analysis of several samples) was: boron 27.1 ± 0.5, phosphorus 72.0 ± 0.5%. This deviates slightly from stoichiometric composition (1:1).

TABLE 5. Determination of Phosphorus after Precipitation by Bismuth Nitrate from an Artificial Mixture of Phosphorus and Boric Acid (P = 18.2, B = 20 mg)

Phosphorus determined, mg		Error of the determination			
from the filtrate	from the residue	from the filtrate		from the residue	
		mg	%	mg	%
18.4	18.3	+ 0.2	1.1	+ 0.1	0.5
18.3	18.1	+ 0.1	0.5	— 0.1	0.5
18.3	18.2	+ 0.1	0.5	0.0	0.5
18.1	18.0	— 0.1	0.5	— 0.2	1.1
18.1	18.1	— 0.1	0.5	— 0.1	0.5
18.1	18.1	— 0.1	0.5	— 0.1	0.5

Powders of ternary B–As–P alloys were obtained by two methods. In the first, a mixture of all three components was taken for the reaction, starting from concentrations lying on the quasi-binary BAs–BP section; in the second, a binary alloy of phosphorus and arsenic was taken and the calculated amount of boron was added. After holding the mixtures in vacuum-sealed quartz ampoules at 800-1100°C, depending on the composition, the phase composition was studied by x-ray diffraction and the proportions of the principal components were determined by chemical analysis. It was found that BAs and BP formed solid solutions of the substitution type (both on the boron phosphide and on the boron arsenide side) up to 30 mol.%. Homogenization of the compositions in the middle of the section failed to produce single-phase samples.

In synthesizing the three-component samples it was found that some arsenic remained left over without reacting; the composition of the resultant powders was therefore determined by chemical analysis. A sample 0.15 g in weight was melted in a platinum crucible with 5-6 g of a mixture of Na_2CO_3, K_2CO_3, and $NaNO_3$ in 1:1:0.25 ratio. The crucible with the mixture of sample and flux was placed in a muffle furnace and the latter was raised to 750°C and held for 10-15 min. The melt was leached with hot water and transferred to a 250 ml measuring flask. In order to determine the boron, 50 ml of solution were neutralized with hydrochloric acid until a rose color was obtained with methyl red, the result was boiled for 5-7 min, cooled, and after installing the glass and calomel electrodes of the LP-58 potentiometer the pH was brought up to 6.9; after introducing mannitol, the solution was titrated with alkali to 6.9 again.

The arsenic was determined bromatometrically after distillation in a Ledebur apparatus [2]. For this purpose 100 ml of solution were evaporated with sulfuric acid in order to remove the nitrogen oxides. The solution was transferred to the distillation apparatus and after adding 1 g of hydrazine sulfate and 0.5 g of potassium bromide the arsenic was distilled in the form of the trichloride. The distillate, heated to 60°C, was titrated with a bromate solution until the red coloring of methyl orange vanished. The phosphorus was determined in the remaining part of the solution. For this purpose the alkaline solution was neutralized with nitric acid by reference to methyl red, and 4 ml of HNO_3 (d = 1.4 g/cm^3) were added to 100 ml of the solution. The total phosphorus and arsenic content was precipitated from the boiling solution by intro-

TABLE 6. Determination of Phosphorus in an Artificial Mixture with Arsenic

Taken, mmole			Determined, mmole		Error of determination	
Arsenic	Phosphorus	Arsenic + phosphorus	Arsenic + phosphorus (indirect – Complexone)	Phosphorus (from the difference)	mmole	%
0.3432	0.5876	0.9308	0.9370	0.5938	+ 0.0062	1.0
0.3432	0.5876	0.9308	0.9232	0.5807	— 0.0069	1.2
0.3432	0.5876	0.9308	0.9254	0.5822	— 0.0054	0.9
0.3432	0.5876	0.9308	0.9272	0.5840	— 0.0036	0.6
0.3391	0.7103	1.0494	1.0247	0.6856	— 0.0247	3.5
0.3391	0.7103	1.0494	1.0280	0.6889	— 0.0214	3.0
0.3391	0.7103	1.0494	1.0369	0.6978	— 0.0125	1.6

ducing a twofold excess of a titrated solution of $Bi(NO_3)_3$ [11]. The residue was filtered through a dense filter and washed with 0.2 N HNO_3.

Then the excess bismuth in the filtrate was titrated with Complexone III, bringing the pH up to 1.0-1.1 with ammonia and adding xylenol orange. The phosphorus was calculated from the difference. Complexonometric titration may also be carried out after dissolving the deposit in hot, dilute (1:1) HNO_3. The results of the precipitation of the total phosphate and arsenate contents from an artificial mixture are shown in Table 6.

The method here proposed enabled ternary powders to be analyzed with completely satisfactory accuracy.

Conclusions

1. We have proposed methods of chemical analysis for compounds of boron with arsenic and phosphorus.
2. On the basis of the chemical analysis, the compound arsenic hexaboride may be assigned the formula B_6As.

Literature Cited

1. Z. Yu. Zndreeva and G. V. Efremov, Vestnik LGU, 2:130-135 (1964).
2. W. F. Hillebrand et al., Practical Handbook on Inorganic Analysis [Russian translation], Goskhimizdat, Moscow (1960).
3. Ya. Kh. Grinberg et al., Izv. Akad. Nauk SSSR, Neorg. Mat., 1:1484 (1965).
4. A. A. Eliseev, A. A. Babitsyna, and É. S. Medvedeva, Zh. Neorg. Khim., 9:1158 (1964).
5. É. S. Medvedeva et al., in: Summaries of Contributions to the Third All-Union Conference on Semiconducting Compounds [in Russian], Kishinev (1963), p. 15.
6. É. S. Medvedeva and G. D. Mitkina, Soviet Patent No. 928,839/23-4 dated from April 3, 1963.
7. A. A. Reshchikova, Zavod. Lab., 31(2):1964 (1965).
8. G. V. Samsonov and Yu. B. Titkov, Zh. Prikladnoi Khim., pp. 363, 669 (1963).
9. N. N. Sentyurina, N. A. Makarova, and É. A. Gerasimova, Zavod. Lab., 29(9):1957 (1963).
10. Yu. V. Shmartsev, Yu. A. Valov, and A. S. Borshchevskii, Refractory Diamond-like Semiconductors [in Russian], Metallurgiya, Moscow (1964).
11. J. Bassett, Analyst, 88(1044):238 (1963).
12. J. Gueilleron and F. Thevenot, Bull. Soc. France, 2:402 (1965).
13. F. V. Williams and R. A. Kuehrwein, J. Am. Chem. Soc., 82:1330 (1960).

COMPLEXONOMETRIC ANALYSIS OF
MOLYBDENUM ALLOYS

L. N. Kugai, O. F. Galadzhii, and V. I. Kornilova

Institute of Problems in Materials Science, Academy of Sciences of the Ukrainian SSR

The most promising method of analyzing alloys containing molybdenum is the complexonometric procedure, which enables the alloys to be analyzed without preliminary separation of the components.

A method has already been described for the complexonometric determination of molybdenum based on the formation of a hexavalent molybdenum complexonate. Molybdenum (VI) and Complexone III form a compound with a molar ratio of Mo:Complexone III = 2:1 at pH 2-7 [1]. As an indicator for the titration one uses pyrocatechin in the presence of an internal light filter of indigocarmine, the pH of the solution being 4-5. Indirect methods of determining hexavalent molybdenum have also been described [5, 12], for example, a method based on the precipitation of calcium molybdate, its dissolution, and subsequent titration of the calcium with a solution of Complexone III in the presence of murexide. More reproducible and accurate results are obtained in the complexonometric determination of molybdenum previously reduced to the pentavalent state.

Molybdenum (V) and Complexone III form a complex with a composition of Mo:Complexone III = 2:1, stable in the acidity range between 0.5 N HCl to pH 10 [2]. The molybdenum is reduced by hydrazine hydrochloride in a boiling solution. The determination of the pentavalent molybdenum is based on the titration of the excess of Complexone III with a solution of a zinc salt in the presence of chromogen black ET as indicator at pH 10.

The complexonometric determination of molybdenum reduced to the pentavalent state may also be carried out with an indicator of 1-(2-pyridyl-azo)-naphthol (PAN) [6]. In this case the molybdenum is reduced with hydroxylamine sulfate, and the excess of the Complexone III is titrated first with a solution of copper sulfate to obtain a bright red color at pH 4.5-5 and then with a solution of Complexone III to a pure yellow color. This kind of titration increases the accuracy of the determination.

Our own problem amounted to the development of a method of complexonometric analysis for ZrB_2–$MoSi_2$, Mo–TiC, Ti–MoC alloys, molybdenum aluminide, and copper–molybdenum alloys without separating the components.

In the complexonometric determination of aluminum the complexonate of the latter is decomposed by sodium fluoride [3, 5, 9, 11] with subsequent determination of the Complexone III released. This method was used in our experiments for analyzing molybdenum aluminide and ZrB_2–$MoSi_2$ alloys.

The principle of the proposed method of analysis lies in the following: First we determine the sum of the zirconium (aluminum) and the molybdenum, previously having reduced the latter to the pentavalent state, by titrating the Complexone III with a solution of zinc chloride in the presence of xylenol orange. To the filtered solution we add ammonium fluoride (to decompose the zirconium complexonate) or sodium fluoride (to decompose the aluminum complexonate) and titrate the Complexone III so released with the same solution of zinc chloride.

In analyzing TiB_2–$MoSi_2$ alloys the titanium is combined with sodium fluoride. In analyzing Mo–Cu alloys the copper is masked with thiourea. The copper is determined by direct titration with Complexone III at pH 8 with a murexide indicator.

Analysis of Molybdenum Aluminides

The sum of the aluminum and molybdenum is determined by titrating the excess of Complexone III with a solution of zinc chloride at pH 9 with an indicator of eriochrome black T. To the filtered solution we add sodium fluoride; this decomposes the aluminum complexonate and the Complexone III released is titrated with a zinc chloride solution.

Analytical Procedure. An alloy sample 0.2-0.3 g in weight is dissolved by heating in 30 ml of sulfuric acid (1:4) with the addition of nitric acid. The solution is evaporated until sulfuric acid vapor appears. In the presence of an insoluble residue this is filtered out, roasted, and fused in potassium pyrosulfate. The solution is transferred to a 100-ml measuring flask.

Determination of Molybdenum + Aluminum. To an aliquot portion of solution (10-20 ml) we add 10 ml of hydrochloric acid (1:1), with a 1% solution of hydrazine hydrochloride, and reduce the molybdenum by boiling the solution for 3-5 min. To the reduced solution we add an excess of Complexone III (0.025 mol. solution), hold for 5-10 min, add eriochrome black T, neutralize with ammonia to a blue-green solution, and titrate the excess of Complexone III with a solution of zinc chloride (0.025 mol. solution) until the solution turns red.

Determination of Aluminum. To the titrated solution we add sodium chloride; we heat the result and titrate the released Complexone III with a solution of zinc chloride (color

TABLE 1. Complexonometric Determination
of Aluminum and Molybdenum in
Artificial Mixtures

Taken, mg		Found, mg		Relative error, %	
Mo	Al	Mo	Al	Mo	Al
Indicator eriochrome black T					
4.70	3.05	4.90	2.93	+ 4.25	− 2.29
7.00	1.53	6.91	1.56	− 1.14	+ 1.92
10.50	2.29	10.30	2.29	− 1.90	0
17.50	5.35	17.23	5.42	− 1.54	+ 1.30
21.02	3.82	21.13	3.89	+ 0.52	+ 1.84
Indicator xylenol orange					
8.17	3.44	8.31	3.49	+ 1.71	+ 1.45
10.04	5.05	9.80	5.05	− 2.30	0
11.67	6.11	11.80	6.11	+ 1.11	0
11.68	7.64	12.04	7.64	+ 3.36	0
12.84	5.35	13.07	5.35	+ 1.79	0
15.17	2.67	15.52	2.67	+ 2.83	0
23.35	3.82	23.67	3.77	+ 1.38	− 1.33

TABLE 2. Reproducibility of the
Determination of the Components in
Molybdenum Aluminides

Taken, mg		Found, mg		Mo/Al ratio	Relative error, %	
Mo	Al	Mo	Al		Mo	Al
21.02	3.82	21.13	3.89	5.5 : 1	+ 0.52	+ 1.84
17.50	5.35	17.23	5.42	3.2 : 1	— 1.54	+ 1.30
11.67	6.11	11.80	6.11	2 : 1	+ 1.11	0
4.70	3.05	4.90	2.98	1.5 : 1	+ 4.25	— 2.29
10.95	18.75	11.58	18.45	1 : 1.7	+ 5.75	— 1.6
13.20	26.42	14.08	26.10	1 : 2	+ 6.66	— 1.21
4.80	12.00	5.40	11 70	1 : 2.5	+ 12.5	— 2.5
6.18	23.70	6.90	23.40	1 : 3.8	+ 10.4	— 1.26

turns from blue-green to red). The amount of Complexone III combined with the aluminum is determined by this titration.

The amount of Complexone III expended on the molybdenum is determined from the difference between the first and second determinations.

In the complexonometric analysis of molybdenum aluminide xylenol orange may also be used as an indicator. In this case the titration is carried out at solution pH 5. The results obtained on titrating artificial mixtures of molybdenum and aluminum are presented in Table 1.

In analyzing molybdenum aluminides it was found that the most accurate and reproducible results were only obtained if the aluminum content of the alloys fell below 50%. Otherwise the conversion of the indicator at the transition point was no longer sharp; the results were inaccurate and poorly reproducible (Table 2).

Analysis of ZrB_2-MoSi_2 Alloys

In these alloys the zirconium is determined by direct titration with Complexone III in a sulfate medium (0.2 N solution) with an indicator of xylenol orange; in another aliquot part of the solution the sum of the zirconium and molybdenum is determined by reverse titration of the excess of Complexone III with a solution of zinc salt and an indicator of xylenol orange. The molybdenum is previously reduced with hydrazine hydrochloride. The zirconium and molybdenum may be determined from a single solution.

At pH 5 we determine the sum of the zirconium and molybdenum (V) by titration with an excess of Complexone III with a solution of zinc salt and xylenol orange. To the filtered solution we add ammonium fluoride in order to decompose the zirconium complexonate, and titrate the Complexone III so released with a solution of zinc chloride and the same indicator. The molybdenum is determined from the difference between the first and second titrations.

Analytical Procedure. An alloy sample 0.2 g in weight is dissolved in a platinum dish by heating in a mixture of HF and HNO_3, then 5-6 ml of sulfuric acid (d = 1.84 g/cm^3) are added and the re-

TABLE 3. Complexonometric
Determination of Molybdenum and
Zirconium in Artificial Mixtures

Taken, mg		Found, mg		Relative error of determination, %	
Mo	Zr	Mo	Zr	Mo	Zr
24.0	20.0	24.2	20.17	+ 0.8	+ 0.85
12.0	20.0	12.07	20.10	+ 0.58	+ 0.5
12.0	10.0	11.90	9.95	— 0.8	— 0.5
18.0	15.0	18.30	14.80	+ 1.66	— 1.33
18.0	25.0	18.32	24.73	+ 1.72	— 1.08

sult is evaporated until sulfuric acid fumes appear. The solution is transferred to a 100-ml measuring flask and water is added to the mark.

Determination of Zirconium + Molybdenum. An aliquot portion of the solution (20 ml) is transferred to a conical flask and the molybdenum is reduced by adding 10 ml of HCl (1:1) and 5 ml of a 1% solution of hydrazine hydrochloride; the result is boiled for 3-5 min till the solution becomes yellow; then an excess of Complexone III is added, held for 5-10 min, five drops of a 0.5% solution of xylenol orange are added, the result is neutralized to pH 5, 10-15 ml of a buffer solution of pH 5 are added, and the excess of Complexone III is titrated with a zinc chloride solution.

Determination of Zirconium. To the titrated solution we add ammonium fluoride and heat the result; we titrate the Complexone III so released with a zinc solution, thus determining the zirconium content. The molybdenum content is found by differencing the first and second titrations.

The zirconium may also be determined in the following manner. To an aliquot portion of the solution (20 ml) we add 200 ml of water and several drops of a solution of xylenol orange, and slowly titrate with Complexone III until the rose color of the solution turns yellow. The method was verified with artificial mixtures. The results are shown in Table 3.

Analysis of Mo–TiC and MoC–Ti Alloys

In analyzing Mo–TiC and MoC–Ti alloys the titanium is masked by means of sodium fluoride. The molybdenum (V) is determined by reverse titration of the excess of Complexone III with a solution of zinc chloride at pH 5 using xylenol orange as indicator, the titanium by reverse titration of the excess of Complexone III with a solution of zinc chloride at pH 5 using the same indicator [8]. When titrating the titanium, the molybdenum (VI) forms a complex with hydrogen peroxide, not reacting with Complexone III.

Analytical Procedure. An alloy sample 0.2 g in weight is dissolved in 20 ml of sulfuric acid (1:4) with the addition of 1-2 ml of nitric acid. The solution is evaporated until sulfuric acid fumes appear, transferred to a 100-ml measuring flask, and brought up to the mark with water.

Determination of Molybdenum. An aliquot portion of the solution (20 ml) is treated with 10 ml of HCl (1:1) and 5 ml of a solution of hydrazine hydrochloride (1% solution) to reduce the molybdenum; then in order to combine the titanium 0.2 g of dry sodium fluoride is added and boiling proceeds for 3-5 min, an excess of a 0.025 M solution of Complexone III is added and held for 5-10 min, five or six drops of a 0.5% solution of xylenol orange are added, the result is neutralized with ammonia to pH 5, 10-15 ml of a buffer solution of pH 5 are added, and the excess of Complexone III is titrated with a solution of zinc chloride until the yellow color of the solution turns orange-red.

Determination of Titanium. To an aliquot portion of the solution (20 ml) we add 2-3 ml of hydrogen peroxide (30% solution), with an excess of Complexone III (0.025 M solution) and hold for 5-10 min; we then add five drops of a 0.5% solution of xylenol orange, neutralize with ammonia to pH 5, add 10-15 ml of a buffer solution (pH 5), and titrate

TABLE 4. Determination of Molybdenum and Titanium in Artificial Mixtures

Taken, mg		Found, mg		Relative error of determination, %	
Mo	Ti	Mo	Ti	Mo	Ti
183.0	55.0	182.00	58.0	− 0.5	+ 5.5
172.0	60.0	173.0	63.0	+ 0.5	+ 5.0
153.0	69.0	155.0	70.0	+ 1.3	+ 1.4
133.0	78.0	132.0	79.0	+ 0.8	+ 1.3
114.0	86.0	114.0	86.0	0	0
95.0	95.0	96.0	96.0	+ 1.0	+ 1.0
76.0	105.0	73.0	103.0	− 3.9	− 1.9
57.0	113.0	55.0	113.0	+ 3.5	0
38.0	121.0	36.0	119.0	− 5.3	− 1.7

TABLE 5. Complexonometric Determination of Titanium in TiC–Mo Samples

Titanium found, %	
with separation of Mo	without separation of Mo
28.7	28.9
24.8	24.8
15.8	15.8
6.9	7.1
4.2	4.2

TABLE 6. Determination of Molybdenum and Copper in Artificial Mixtures

Taken, mg		Found, mg		Relative error of determination, %	
Mo	Cu	Mo	Cu	Mo	Cu
13.25	10.0	13.0	10.0	− 1.88	0
15.00	12.0	15.25	12.15	+ 1.66	+ 1.25
15.00	8.00	15.50	8.00	+ 3.33	0
31.25	5.5	31.50	5.45	+ 0.8	− 0.9
11.00	10.0	11.25	10.10	+ 2.27	+ 1.0
16.25	15.00	16.10	15.05	− 0.92	0
16.25	8.0	16.45	7.84	+ 1.23	− 2.00
17.50	12.80	18.00	12.48	+ 2.20	− 2.5
18.75	8.00	19.00	8.00	+ 1.33	0
25.00	9.60	24.65	9.60	+ 1.40	0

the excess of Complexone III with a solution of zinc chloride until the color of the solution changes from yellow to orange-red.

This method of analysis was verified for artificial mixtures and alloy samples. In analyzing TiC–Mo alloys, the titanium and molybdenum were separated by precipitating the molybdenum with Cupferron. The results are shown in Tables 4 and 5.

Analysis of Molybdenum–Copper Alloys

In analyzing molybdenum–copper alloys the copper is masked with thiourea. The copper is determined by titration with Complexone III at pH 8 using a murexide indicator; the molybdenum, after previously reducing it and combining the copper with thiourea, by reverse titration of the excess of Complexone III with a solution of zinc chloride at pH 5, using an indicator of xylenol orange.

Analytical Procedure. A sample 0.3 g in weight is dissolved by heating in 15–30 ml of nitric acid (1:1), 10 ml of sulfuric acid (1:1) are added, and the result is evaporated until sulfuric acid fumes appear.

The solution is transferred to a 100-ml measuring flask. In order to determine the molybdenum an aliquot portion of solution (20 ml) is transferred to a conical flask, 10 ml of HCl (1:1) are added with 5 ml of a hydrochloric acid solution of hydrazine, and the result is boiled for 3–5 min until it becomes yellow. The hot solution is then treated with 0.5 g of thiourea and an excess of Complexone III (held for 5–10 min, neutralized with ammonia to pH 5) and 10–15 ml of a buffer solution (pH 5) are added, with eight or ten drops of xylenol orange (0.5% solution); titration then proceeds with zinc chloride until the yellow color turns orange-red.

In order to determine the copper, an aliquot portion of solution (20 ml) is transferred to a conical flask, diluted with water to 50–70 ml, neutralized with ammonia to pH 8, and titrated with Complexone III, using murexide as indicator.

The method was verified for artificial mixtures. The results appear in Table 6.

Conclusions

We have developed methods of complexonometrically analyzing molybdenum aluminides, ZrB_2–$MoSi_2$, MoC–Ti, TiC–Mo alloys, and copper–molybdenum alloys without preliminary separation of the components.

Literature Cited

1. A. I. Busev and Fan' Chzhan', Vestnik MGU, Ser. Mat., Mekh., Astron., Fiz., i Khim., 2:203 (1959).
2. A. I. Busev and Fan' Chzhan', Zh. Analit. Khim., 14:445 (1959).
3. A. I. Busev, A. G. Petrenko, and I. A. Bykovskaya, Zavod. Lab., 27(6):659-661 (1961).
4. S. V. Elinson and L. I. Pobedina, Zavod. Lab., 29(2):139 (1963).
5. F. Lassner and H. Schlesinger, Z. analyt. Chem., 158:195 (1957).
6. E. Lassner and R. Scharf, Z. analyt. Chem., 167(2): 114-117 (1959).
7. R. Pribil and V. Vesely, Talanta, 10(Feb.):233-235 (1963); Referat. Zh. Khim., 14G85 (1963).
8. R. Pribil and V. Vesely, Talanta, 10(Apr.):383-386' (1963); Referat. Zh. Khim., 23G30 (1963).
9. Istvan. Sajo, Magyar kém. folyóirat, 60(9):268 (1954).
10. Istvan. Sajo, Magyar kém. folyóirat, 62(2):56-59 (1956).
11. Istvan. Sajo, Acta Chim. Sci. Hung., 61(3-4):233 (1955).
12. A. Sousa, Analyt. Chim. Acta, 12:215 (1955).
13. Donald H. Wilkins, Referat. Zh. Khim., 15:53138 (1959).

ANALYSIS OF TITANIUM, ZIRCONIUM, HAFNIUM, AND TANTALUM GERMANIDES

G. T. Kabannik and O. I. Popova

Institute of Problems in Materials Science, Academy of Sciences of the Ukrainian SSR

Large quantities of germanium are determined by gravimetric and volumetric methods. Gravimetric methods are based on the precipitation of germanium with various precipitating agents, for example, the formation of a salt of pyrocatechin-germanic acid with o-phenanthroline [6], hydroxyquinoline [2], and others. The volumetric method is based on the capacity of germanic acid to form a complex acid with polyatomic alcohols or monosaccharides, this being titrated with alkali using various indicators [3, 8].

Germanium is separated from the majority of interfering elements by the distillation of germanium chloride [1, 10] or extraction with organic solvents [3, 9], precipitation by hydrogen sulfide, etc.

In order to decompose samples containing germanium, one employs melting with alkali and sodium peroxide, sintering with a mixture of calcium oxide and potassium nitrate [4], or sintering with calcium oxide and magnesium nitrate. Dissolution in a mixture of sulfuric, hydrofluoric, and phosphoric acids, etc., is also recommended [7, 12].

We desired to analyze the germanides of various transition metals: titanium, zirconium, hafnium, and tantalum, but found no published data relating to the required procedure.

The germanides of the transition metals are chemically stable compounds. Titanium, zirconium, and hafnium germanides are dissolved in a mixture of HF and HNO_3 and sintered with various mixtures such as barium carbonate + calcium oxide, magnesium oxide + sodium carbonate, etc. Titanium germanide is also completely dissolved in a 30% solution of hydrogen peroxide. Tantalum germanide is dissolved in a mixture of potassium sulfate and sulfuric acid.

These results show that the choice of reagents for bringing germanides into solution is extremely limited if one considers that on sintering (and also on melting) with alkali or sodium peroxide in nickel or iron crucibles additional foreign components are introduced into solution.

Analysis of Titanium, Zirconium, and Hafnium Germanides

For decomposing these compounds we used the method of dissolution in a mixture of HF and HNO_3 and further evaporation with sulfuric acid until the vapor of the latter appeared.

Titanium, zirconium, and hafnium were determined complexonometrically (by reverse titration with an excess of Complexone). It is well known [5] that prolonged heating is needed

54

for the complexonometric determination of germanium. Our own experiments showed that germanium entirely failed to interact with Complexone III in the cold under the conditions used for determining titanium, zirconium, and hafnium, and accordingly in no way interfered with their determination.

In order to determine the germanium we chose a titrometric method, titration of a complex mannitogermanic acid with alkali. First we had to study methods of eliminating the perturbing influence of the cation (Ti, Zr, Hf). Separation with barium carbonate was undesirable, firstly because the deposits absorbed a certain amount of germanium, and secondly because the method was unsuitable in the presence of sulfuric acid.

In determining germanium we tried using the method of masking the cation with complexing agents and achieved successful results. We tried tartaric acid, Rochelle salt, and hydrogen peroxide (the latter only for titanium).

It is well known [11] that germanium forms complexes with tartaric acid over a wide range of pH values with a 1:1 ratio of the components; these are not stable and are decomposed by mannitol; hence mannitogermanic acid is stable.

Analytical Procedure

A germanide sample 0.1-0.2 g in weight was dissolved in a mixture of HF and HNO_3, 20 ml of sulfuric acid (1:1) were added, and the result was evaporated by heating slowly until sulfuric acid fumes appeared. On cooling, the solutions were transferred to measuring flasks and diluted with water up to the mark. If a white precipitate of germanic acid was formed during the evaporation, this was filtered off, washed twice with water, dissolved in 5-10 ml of a hot 20% solution of caustic soda, and added to the filtrate in the measuring flask. Titanium germanide may also be dissolved in 30% hydrogen peroxide.

In order to determine titanium, an aliquot portion of the solution is treated with 1 ml of 30% hydrogen peroxide and an excess of a titrated solution of Complexone III; after 10 min the solution is neutralized to pH 5, an ammonium acetate buffer solution of pH 5 is added, and the excess of Complexone III is titrated with a solution of zinc sulfate using an indicator of xylenol orange. The titer of the Complexone III with respect to the elements is determined by reference to standard solutions of the corresponding metal under conditions similar to those employed for the determination.

Hafnium and zirconium are determined analogously, although without the hydrogen peroxide and 10-min holding period. In order to determine the germanium, an aliquot portion of the solution is treated with a solution of tartaric acid (for hafnium and zirconium), 0.02-0.03 g to 0.02 g of germanide, or two to four drops of hydrogen peroxide (for titanium), this quantity also relating to 0.02 g of the germanide. Then the solution is neutralized with respect to phenolphthalein with a 20% solution of caustic soda, and one drop of sulfuric acid (1:4) is added. The excess of the latter is neutralized with a titrated solution of 0.02 N alkali until a slightly red color of phenolphthalein appears. Then an excess of mannitol is introduced, and after 5 min the mannitogermanic acid so formed is titrated with the same alkali. The titer of the alkali with respect to germanium is established by reference to a standard solution of germanium under the same conditions as those characterizing the determination.

Analysis of Tantalum Germanide

A sample of tantalum germanide may be dissolved either in an $HF-HNO_3$ mixture, with the subsequent addition of H_2SO_4 and evaporation until the fumes of the latter appear, or else by wet fusion with a mixture of potassium sulfate and sulfuric acid. Subsequently the resultant solution is distilled in the presence of a large amount of HCl (final concentration 6 N), and the

TABLE 1. Determination of Germanium in Artificial Mixtures

System	Metal taken, mg	Germanium taken, mg	Germanium found, mg	Relative error, %
Ge — Zr	5.60	6.90	6.74	— 2.3
Ge — Zr	5.60	13.81	14.10	+ 2.1
Ge — Ti	3.60	12.77	12.50	— 2.1
Ge — Ti	3.60	5.52	5.69	+ 3.0

TABLE 2. Determination of Germanium in Various Samples by the Method of Addition

Germanide	Ge content, %	Ge taken, mg	Ge introduced, mg	Total Ge content, mg	Ge found, mg	Relative error, %
$ZrGe_2$	57.7	5.69	12.77	18.46	18.22	— 1.2
		5.69	12.77	18.46	18.13	— 1.7
		5.52	6.38	11.9	12.03	+ 1.5
Zr_5Ge_3	32.3	5.52	6.90	12.42	12.26	— 1.2
		5.52	6.90	12.42	12.77	+ 2.8
Ti_5Ge_3	47.5	8.11	6.38	14.49	14.49	0
		8.11	6.38	14.49	14.67	+ 1.2
Hf_5Ge_3	17.8	7.59	6.90	14.49	14.49	0
		7.59	6.90	14.49	14.16	— 2.3

germanium is determined in the receiver (in the distillate) by the titration of the germanium–mannitol complex with alkali.

In the solution remaining in the reaction vessel, the tantalum is determined by precipitation with ammonia and subsequent weighing in the form of tantalum pentoxide. In analyzing tantalum germanide, the germanium may clearly be separated by extraction of the chloride.

Tables 1 and 2 show the results of experiments relating to the determination of germanium in artificial mixtures and by the method of additions.

Conclusions

We have developed methods of analyzing the germanides of titanium, zirconium, hafnium, and tantalum. In order to determine the germanium, the alkalimetric method of titrating mannitogermanic acid is employed. The interfering effects of titanium, zirconium, and hafnium in the determination of germanium are eliminated by the introduction of tartaric acid (for Hf and Zr) or hydrogen peroxide (for Ti). The titanium, zirconium, and hafnium are titrated complexonometrically in the presence of germanium.

Literature Cited

1. I. P. Alimarin, Trudy Vses. Konf. Analit. Khim., 2:371 (1943).
2. I. P. Alimarin and O. A. Alekseeva, Zh. Prikladnoi Khim., 12:1900 (1939).
3. S. D. Gur'ev and N. N. Lutchenko, Trudy Gos. NII Tsvet. Metallov, 19:722 (1922).
4. S. A. Dekhtrikyan, Dokl. Akad. Nauk Arm. SSR, 28:213 (1959).
5. V. A. Nazarenko et al., Zh. Analit. Khim., 19:787 (1964).
6. V. A. Nazarenko and A. M. Andrianov, Zavod. Lab., 29:795 (1963).
7. V. A. Nazarenko, N. A. Lebedeva, and R. V. Ravitskaya, Zavod. Lab., 24:9 (1958).
8. H. I. Gluley, Analyst, 76:517 (1951).
9. W. Fischer and W. Haare, Angew. Chem., 66:165 (1954).
10. W. Fischer and H. Kleim, Z. analyt.Chem., 128:443 (1948).
11. D. A. Everest and Harrisson, J. Amer. Chem. Soc., p. 3752 (1960).
12. E. H.Strickland, Analyst, 80:548 (1955).

SOME DATA RELATING TO THE
CHEMICAL PROPERTIES OF GERMANIDES

O. I. Popova

Institute of Problems in Materials Science, Academy of Sciences of the Ukrainian SSR

Published data relating to the chemical properties of metal germanides are very limited. The germanides of the alkali metals have been studied most completely. Thus lithium germanides of the following compositions are known: Li_4Ge and Li_6Ge_2. These decompose in damp air, forming Li_2CO_3; Li_4Ge is more active [1].

Hohmann and Johnson [6, 9] described the properties of potassium, sodium, rubidium, and cesium germanides. Alkali germanides of the following compositions were synthesized: KGe, NaGe, RbGe, CsGe, and the germanium-enriched compounds KGe_4, $NaGe_4$, $RbGe_4$, $CsGe_4$. All these were only stable in an atmosphere of dry inert gas. In air they decompose, forming the corresponding alkali and a brown, amorphous substance, germanium hydride $(GeH)_x$.* The latter is oxidized by various oxidizing agents (nitric acid, hydrogen peroxide, and so on) to germanium dioxide. When the compounds of the Na–Ge system interact with water or hydrochloric acid, germanium hydride is also formed. On heating in high vacuum to 480°C the germanium-sodium compound dissociates completely.

Magnesium forms a germanide Mg_2Ge of silver-gray color. Like the germanides of the alkali metals, this compound is unstable in air and after 5 min decomposes with the formation of germanium hydride; it is readily decomposed by water [4].

Calcium forms two compounds with germanium: GeCa (silver-gray color) and $GeCa_2$. Both are unstable in air and decompose with the formation of $Ca(OH)_2$ or $CaCO_3$ and an orange powder [7, 8], apparently germanium hydride.

The copper germanides Cu_3Ge and Cu_5Ge are very stable toward HCl and H_2SO_4. Copper alloys containing over 25% of germanium only decompose under the action of aqua regia [3]; on reducing the germanium content, HNO_3 suffices.

Thorium and germanium form $ThGe_3$ and the α and β forms of $ThGe_2$. These compounds interact vigorously at room temperature with 50% HCl, concentrated HF, aqua regia, and a 10% solution of NaOH, and slowly with 30% hydrogen peroxide, 3 N and 18 N sulfuric acid, and concentrated and 6 N nitric acid, not reacting at all with 85% phosphoric acid or a 0.1 N solution of potassium permanganate [2].

---

*We shall not present any data relating to the chemical properties of the germanides of the rare-earth elements.

TABLE 1. Chemical Stability of Titanium, Hafnium, and
Zirconium Germanides

Solvent	Insoluble residue, %		
	Ti_5Ge_3	Hf_5Ge_3	Zr_5Ge_3
Conc. hydrochloric acid	19	21	17
Hydrochloric acid (1:1)	20	15	15
Nitric acid (conc.)	*	84	–
Nitric acid (1:1)	*	88	79
Sulfuric acid (conc.)	19	10	12
Sulfuric acid (1:4)	37	66	54
Nitric acid + hydrochloric acid (1:3)	0	0	0
Nitric acid + hydrofluoric acid	0	0	0
Hydrochloric acid + hydrogen peroxide	17	17	13
Hydrogen peroxide (30%)	0	99.5	–
Bromine water	99.5	100.0	–
Sulfuric acid (1:4) + 25% solution of ammonium persulfate	12	73	–

* Ti_5Ge_3 decomposes completely, forming a hydrolytic residue.

Of all the germanides of the vanadium subgroup, only the chemical properties of the niobium germanides have been described in the literature [5]. The niobium germanides $NbGe_2$ and Nb_3Ge react energetically with molten soda and alkali, 48% HF, 30% hydrogen peroxide, and a cold NaOH solution. Both germanides react with concentrated sulfuric acid on heating, but not with aqua regia, dilute or concentrated HCl and HNO_3, or 6 N sulfuric acid.

Molybdenum and germanium form several germanides: Mo_3Ge, Mo_3Ge_2, Mo_2Ge_3, $MoGe_2$. All these react with alkalis, are dissolved by hydrogen peroxide, aqua regia, and mixtures of nitric and hydrofluoric acids [11]. The germanides Mo_2Ge_3 and $MoGe_2$ are dissolved in nitric acid or hydrogen peroxide, forming a deposit of germanium dioxide GeO_2.

Iron forms the germanides $FeGe_2$ and Fe_2Ge; increasing the germanium content in the Fe—Ge system increases the corrosion resistance in air; air fails to change $FeGe_2$ at all.

The chemical properties of the following manganese germanides have been studied: Mn_3Ge_2, Mn_5Ge_3, Mn_5Ge_2. The Ge-rich compounds are unstable in acids and alkalis [1].

The chemical stability of rhenium germanide ($ReGe_2$) has been studied in the cold (15 min-24 h) and on boiling (5 min) [10]. The experiments showed that rhenium germanide was very resistant to chemical reagents. Thus the compound failed to decompose in concentrated HCl, cold concentrated and dilute H_2SO_4, concentrated and dilute HNO_3, 85% H_3PO_4, 68% $HClO_4$, a 0.1 N solution of potassium permangante, 30% hydrogen peroxide, molten soda, and alkali. Only hot concentrated sulfuric acid and caustic soda decompose this germanide.

We see from the foregoing that the chemical properties of many germanides have never been studied. We accordingly studied the solubility of titanium, zirconium, hafnium, tantalum, chromium, and molybdenum germanides in various solvents. A sample of germanide (1 g) was placed in a 100-ml beaker and various solvents or mixtures of these were poured over it. The beakers were covered with watch glasses and the contents were boiled for 1 h. In order to keep the volume and concentration of the solution at a specified level, volatile or decomposing solutions were replenished a number of times. On cooling the solution, the insoluble residue was filtered into a crucible (Schott No. 4) and weighed.

TABLE 2. Chemical Stability of Tantalum Germanide

Solvent	Insoluble residue, %
Hydrochloric acid (conc.)	97.0
Hydrochloric acid (1:1)	98.0
Sulfuric acid (conc.)	70.0
Sulfuric acid (1:1)	99.0
Sulfuric acid (1:4)	100.0
Nitric acid (conc.)	98.0
Nitric acid + hydrochloric acid (1:3)	99.0
Nitric acid + hydrofluoric acid	0
Hydrochloric acid + hydrogen peroxide	99.0
Hydrogen peroxide (30%)	97.0
Acetic acid	98.0
Sulfuric acid + potassium sulfate	0
Caustic soda (1%)	88.0
Phosphoric acid	99.0
Bromine water	98.0
Ammonia + hydrogen peroxide	97.0
Ammonium fluoride + hydrogen peroxide	53.0
Caustic soda + hydrogen peroxide	54.0
Ammonium fluoride + ammonium persulfate	37.0

TABLE 3. Chemical Stability of Chromium and Molybdenum Germanides

Solvent	Insoluble residue, %			
	Cr_3Ge	$MoGe_2$	Mo_3Ge	Mo_2Ge_3
Sulfuric acid (conc.)	47[*]	–	–	–
Sulfuric acid (1:4)	99	–	–	–
Sulfuric acid (1:1)	51	–	74	–
Hydrochloric acid (conc.)	0	96	97	97
Hydrochloric acid (1:1)	–	100	98	97
Nitric acid (1:1)	–	0	0	0
Phosphoric acid	–	–	100	98
Acetic acid	–	53	98	93
Hydrochloric acid + nitric acid (3:1)	15	0	0	0
Sulfuric acid + hydrochloric acid	6	–	–	–
Hydrogen peroxide	–	–	100	0
Ammonia + hydrogen peroxide	–	–	25	39
Hydrochloric acid + hydrogen peroxide	0	5	20	12
Sulfuric acid + hydrofluoric acid	0	–	–	–
Bromine water	–	29	19	21
Hydrochloric acid + ammonium persulfate	–	40	–	35
Caustic soda (20%)	–	–	99	–
Caustic soda + bromine water	74	–	–	–

[*]Formation of salts.

60   O. I. POPOVA

We see from the results of Table 1 that all the germanides of the titanium subgroup Ti_5Ge_3, Hf_5Ge_3, Zr_5Ge_3 are completely decomposed by an $HF-HNO_3$ mixture; concentrated and dilute HCl and H_2SO_4 dissolve them to a considerable extent, and aqua regia completely.

The compound Ti_5Ge_3 completely decomposes in 30% hydrogen peroxide and nitric acid.

Tantalum germanide Ta_5Ge_3 is quite stable toward ordinary chemical reagents (Table 2); it is not dissolved in hydrochloric or sulfuric acids or in aqua regia, only being completely dissolved by wet fusion in a mixture of potassium sulfate and sulfuric acid or in mixtures of nitric and hydrofluoric acids, and to a considerable extent in mixtures of caustic soda and hydrogen peroxide, hydrogen peroxide and ammonium fluoride, or ammonium fluoride and ammonium persulfate.

Table 3 illustrates the solubility of chromium and molybdenum germanides. We see from these data that Cr_3Ge is unstable with respect to ordinary acids; it dissolves in hydrochloric and sulfuric acids or a mixture of sulfuric and hydrofluoric acids, being fairly soluble in concentrated sulfuric acid, with the simultaneous formation of basic salts. In a mixture of potassium sulfate and sulfuric acid, this compound dissolves almost completely, although it also involves the formation of basic salts. Nitric acid and mixtures of this with other acids (e.g., sulfuric) fail to dissolve chromium germanide owing to the passivation of the chromium.

The results of our study of the solubility of germanides in acids in general support existing data; it should be noted, however, that in the dissolution of Mo_2Ge_3 and $MoGe_2$ in nitric acid and hydrogen peroxide we failed to detect any formation of a deposit of germanium dioxide.

The germanide $MoGe_2$ partly dissolves in acetic acid. All the molybdenum germanides undergo considerable dissolution in bromine water, and $MoGe_2$ and Mo_2Ge_3 in a mixture of hydrochloric acid with ammonium persulfate.

Literature Cited

1. A. E. Vol, Compounds and Properties of Binary Metallic Systems [in Russian], Fizmatgiz, Moscow (1962).
2. H. Nowotny, Uspekhi Khimii, 27:996 (1958).
3. O. Schwartz, Uspekhi Khimii, 5:1448 (1936).
4. G. Brauer and J. Tiesler, Z. anorg. Chem., 262:319 (1950).
5. J. H. Carpenter and A. W. Searcy, J. Amer. Chem. Soc., 78:2079 (1956).
6. E. Hohmann, Z. anorg. Chem., 257:113 (1948).
7. P. Eckerlin, H. J. Meyer, and E. Wölfel, Z. anorg. Chem., 281:322 (1955).
8. P. Eckerlin and E. Wölfel, Z. anorg. Chem., 280:321 (1955).
9. O. H. Johnson, Chem. Rev., 51:431 (1952).
10. A. W. Searcy, R. A. McNeels, and J. A. Cristions, J. Amer. Chem. Soc., 76:5287 (1954).
11. A. W. Searcy and R. J. Reavler, J. Amer. Chem. Soc., 75:5657 (1953).

ANALYSIS OF THE SELENIDES OF
THE RARE-EARTH ELEMENTS

V. A. Obolonchik and T. M. Mikhlina

Institute of Problems in Materials Science, Academy of Sciences of the Ukrainian SSR

The selenides of the rare-earth metals have mainly been studied in connection with their physical properties. Nothing has been published regarding methods of chemical analysis, only data obtained by x-ray structural and x-ray phase analysis being generally mentioned.

A preliminary study of the chemical properties of the selenides of the rare-earth elements enabled us to develop some methods of chemical analysis. In studying the solubility of these selenides we came to the conclusion that it was most reasonable to determine the total amount of selenium and metal in a nitrate solution. Nitric acid completely dissolves the rare-earth selenides with the formation of selenious acid and metal nitrates. The determination of selenium in selenious acid presents no difficulties. Methods of determining selenium in solution in the form of SeO_3^{2-} have been described in the literature [2]. In developing a method of determining free selenium we studied the solubility of the selenides in water, and then used the reaction based on the formation of a selenosulfate by interaction between selenium and sodium sulfite:

$$Na_2SO_3 + Se \rightleftarrows Na_2SeSO_3.$$

The constant of the reaction depends on the temperature [1]:

$$K = \frac{[Na_2SeSO_3]}{[Na_2SO_3]}.$$

With increasing temperature the reaction moves in the direction of the formation of the selenosulfate. This is stable in a weakly alkaline or neutral medium in the presence of an excess of sulfite, which moves the equilibrium to the left:

$$SeSO_3^{2-} \rightleftarrows SO_3^{2-} + Se.$$

The removal of the sulfite from solution with formaldehyde or the acidification of the solution leads to a quantitative precipitation of the red form of selenium:

$$SeSO_3^{2-} + H^+ \rightarrow Se + HSO_3^-.$$

TABLE 1. Verification of the Amount of
Free Selenium in the Cerium Selenide
by the Method of Additions

Weight of Ce$_2$Se$_3$ sample, mg	Se free content, mg	Added Se, mg	Se content, mg		Relative error, %
			expected	found	
110	0.44	15	15.44	15.65	+1.35
115	0.46	22	22.46	22.81	+1.56
112	0.44	17	17.44	17.67	+1.30
105	0.42	14	14.42	14.67	+1.75
108	0.43	16	16.43	16.25	−1.10

In an acid medium bromine oxidizes the selenosulfate to selenious acid and sulfate. The latter in no way interferes with the determination of selenium. The analytical method was verified by reference to the method of additions (Table 1). The relative error in the determination was about 2%.

In this way, by using the methods of chemical analysis (volumetric and gravimetric) we may determine the free and total selenium contents and the proportion of rare-earth metal present.

In order to determine the total selenium content the selenides are decomposed by nitric acid, which oxidizes all the selenium, free and combined, to selenious acid. Since the reaction takes place quite vigorously, the experiment should be conducted in a conical flask with a reflux air condenser in order to avoid loss of selenium in the dissolution, the walls of the condenser being moistened with acid. For the gravimetric method the selenious acid is reduced by SO$_2$ gas to elemental selenium; for the volumetric method we use the interaction of selenious acid with potassium iodide; the iodine separating is titrated with sodium thiosulfate, using starch as an indicator. In order to ensure that the titration should not be impeded by precipitating red amorphous selenium, which adsorbs iodine, the starch solution should be added before the iodide. In the resultant colloidal solution of selenium the blue color turns orange very sharply.

In determining the free selenium content, a sample of selenide is heated together with a 2 N solution of sodium sulfite in a flask with a reflux condenser for 40 min, using a boiling water bath. The selenium is determined in the filtrate either gravimetrically or volumetrically.

The gravimetric method is based on the decomposition of the selenosulfate with the release of elemental selenium. To the solution of selenosulfate we add formalin; then the removal of the excess of sulfite from solution by formaldehyde leads to a quantitative precipitation of selenium.

The volumetric method is based on the oxidation of the selenosulfate with bromine water to selenious acid. The excess of bromine is removed by slight boiling with the subsequent addition of several drops of an alcohol solution of acetanilide. The selenium in the selenious acid is determined in the manner indicated earlier.

The determination of the rare-earth metals in the selenides is based on their decomposition with nitric acid and subsequent titration with a solution of Complexone III. The lanthanide ion forms a stable complex with this reagent. The selenious acid in solution causes no interference. An aliquot portion of the solution is neutralized with ammonia to pH 4.5, an acetate buffer solution is added together with a mixed indicator of Alizarin C and methylene blue, and titration with Complexone III proceeds while boiling until the color turns from crimson red to a dirty green.

In view of the fact that the original materials, the rare-earth oxides, contain traces of Fe, Ca, and Cu, it is particularly interesting to test the amounts of the latter in the selenides. It has been shown by some authors that iron selenide sublimes at temperatures above 900°C. The iron in the oxides is determined as specified in the All-Union State Standard, but in order to determine the iron in the selenides a method based on the colorimetric determination of tervalent iron with sulfosalicylic acid or divalent with o-phenanthroline has been developed. The tervalent iron may be reduced to divalent by hydroxylamine hydrochloride or sulfur dioxide gas, i.e., the same solution as that used for the determination of selenium by the gravimetric procedure may be employed.

The results of the analysis showed that in selenides obtained at temperatures of 1100°C the iron content was only half that in the oxides, while in the case of 800°C it was much the same as the latter. This confirms the possibility of obtaining compounds with fewer impurities than the original materials by high-temperature synthesis.

Literature Cited

1.   S. M. Golyand, Khimicheskaya Prom., No. 2 (1947).
2.   M. I. Troitskaya, Methods of Determining and Analyzing the Rare Elements [in Russian], Izd. AN SSSR, Moscow (1961), p. 580.

COMPLEXONOMETRIC ANALYSIS OF ALLOYS OF THE RARE-EARTH OXIDES WITH THE OXIDES OF GROUP II ELEMENTS AND CHROMIUM

S. F. Boremskaya and G. T. Kabannik

Institute of Problems in Materials Science, Academy of Sciences of the Ukrainian SSR

One of the main problems now facing chemists is the creation of new materials suitable for working at high temperatures. Considerable interest in this connection is presented by the oxides of the rare-earth elements, yttrium, and scandium, and also the compounds and solid solutions which these form with the refractory oxides of group II elements, since these compounds have extremely high melting points (about 2000°C) and other valuable properties.

The character of the interactions between the oxides of the rare-earth elements, yttrium, and scandium and the refractory oxides of the group II elements (magnesium, calcium, strontium, and barium oxides) has hardly been studied at all. Such an investigation is therefore of both theoretical and practical interest for the creation of new refractory materials on the basis of these oxides and the production of special glasses and electrical ceramics. The alloys of the rare-earth elements with chromium oxides also possess interesting and valuable properties.

In the analysis of a large number of samples, it is particularly important to use methods yielding reasonably accurate determinations with the minimum expenditure of time. Such methods include complexonometric titration, which is widely employed in laboratory practice.

In order to determine rare-earth elements and separate them from others the gravimetric oxalate method has until recently been the most frequently employed; however, this is inconvenient for routine work in view of the considerable time taken and the relatively low accuracy.

The most reasonable is the volume-complexonometric method. At the present time several complexonometric methods of determining rare-earth elements forming fairly stable complexes with Complexone III have been published, both direct and reverse titration being envisaged.

In order to determine rare-earth elements of the yttrium subgroup the titration of Complexone III in a weakly acid medium with a mixed indicator (ammonium salt of alizarinsulfoacid with methylene blue) has been recommended [4]. There is also the method of titrating rare-earth elements with an indicator of eriochrome black T [5]. In order to create a specified acidity, urotropin is employed (pH 5). As an indicator in the complexonometric titration of rare-earth elements it is recommended to use xylenol orange [8] and arsenazo [7]. Titration with an indicator of arsenazo is recommended at pH 5.5-6.5 in the presence of pyridine.

TABLE 1. Complexonometric Determination of Rare-Earth
and Alkaline-Earth Elements in Artificial Mixtures

Ytterbium oxide–strontium oxide						Ytterbium oxide–barium oxide					
Ytterbium			Strontium			Ytterbium			Barium		
Taken, mg	Found, mg	Relative error, %	Taken, mg	Found, mg	Relative error, %	Taken, mg	Found, mg	Relative error, %	Taken, mg	Found, mg	Relative error, %
15,23	15.23	0	18.48	18.61	+0.72	21.76	21.76	0	13.75	13.84	+0.67
20.4	20.4	0	16.63	16.63	0	4 .8	41.07	+0.67	20.17	20.17	0
10.88	11.1	+1.2	13.2	13.2	0	17.68	17.54	−0.76	12.84	13.02	+1.4
10.88	11.01	+1.2	10.56	10.56	0	10.88	10.88	0	13.66	13.73	+1.0
54.4	54.4	0	27.72	27.72	0	40.8	40.93	+0.33	27.2	27.2	0
38.08	38.08	0	44.88	45.14	+0.6	48.96	49.23	+0.56	38.08	37.8	−0.71
29.92	30.19	+0.9	28.51	28.38	−0.46	54.4	54.4	0	44.88	44.88	0
48.96	49.23	+0.55	62.04	62.17	+0.2	108.8	108.8	0	68.00	68.00	0

Yttrium oxide–calcium oxide					
Yttrium			Calcium		
Taken, mg	Found, mg	Relative error, %	Taken, mg	Found, mg	Relative error, %
40.1	40.5	+1.0	76.8	76.5	0.42
46.1	46.3	+0.44	65.4	65.1	−0.15
80.2	80.6	+0.5	98.1	99.7	+1.67
60.2	60.6	+0.67	65.4	64.9	−0.75
122.3	122.3	0	32.4	32.7	+1.0
75.0	75.0	0	50.6	50.6	0
46.1	46.1	0	32.7	32.7	0
69.8	70.2	+0.57	58.8	58.8	0

Yttrium oxide–strontium oxide						Yttrium oxide–barium oxide					
Yttrium			Strontium			Yttrium			Barium		
Taken mg	Found, mg	Relative error, %	Taken, mg	Found, mg	Relative error, %	Taken, mg	Found, mg	Relative error, %	Taken, mg	Found, mg	Relative error, %
60.1	60.9	+1.3	26.3	25.9	−2.0	60.2	60.5	+0.67	18.3	18.1	−1.0
80.2	80.2	0	39.5	39.5	0	40.1	40.1	0	21.9	21.8	−0.83
80.2	80.2	0	7.9	8.0	+0.17	80.2	80.8	+0.75	14.6	14.4	−1.25
46.5	46.1	−0.86	26.4	26.6	+1.0	70.2	70.5	+0.57	21.0	20.9	−0.87
56.1	56.1	0	47.5	47.5	0	41.7	42.1	+0.96	36.6	36.4	−0,5
46.1	46.1	0	38.0	38.2	+0.7	56.9	56.9	0	11.3	11.4	+0.8
66.1	65.8	−0.6	52.7	52.9	+0.25	98.2	98.2	0	18.3	18.3	0
160.4	160.4	0	40.9	40.9	0	200.5	200.5	0	36.6	36.7	+0.25

Several versions of the determination of rare-earth elements by reverse titration have been proposed, for example, the titration of an excess of Complexone III with a solution of zinc chloride or sulfate with an indicator of eriochrome black T [6].

S. P. Onosova [2] proposes a method of the reverse titration of an excess of Complexone III with a solution of nickel chloride at pH 10 in the presence of murexide; Pribil [9] proposed a method of reverse titration of an excess of Complexone III with lead nitrate at pH 5.5, using a xylenol orange indicator.

The ability of the rare-earth elements, scandium, and yttrium to form complexonates at pH 3-7 and magnesium and calcium at pH 10 [1] was used in developing a method for the complexonometric analysis of the oxide alloys of the rare-earth elements, scandium, and yttrium with alkaline-earth elements.

TABLE 2. Determination of Rare-Earth
and Alkaline-Earth Elements (wt.%)
by the Gravimetric and Complexonometric Methods

Sample	Rare-earth elements		Alkaline-earth elements	
	gravimetric oxalate method	complexonometric method	gravimetric oxalate method	complexonometric method
Y_2O_3 — SrO	67.15	67.01	32.8	32.76
Yb_2O_3 — SrO	77.85	77.8	21.9	21.9
Y_2O_3 — BaO	57.5	57.5	42.6	42.5
YB_2O_3 — BaO	70.45	70.4	29.55	29.5
Y_2O_3 — CaO	76.83	76.7	23.2	23.1
Y_2O_3 — CuO	10.05	9.9	89.9	89.8
Y_2O_3 — CuO	99.2	99.1	0.7	0.7
Y_2O_3 — CaO	6.4	6.3	93.5	93.6
Yb_2O_3 — MgO	74.0	74.1	25.8	25.9
Yb_2O_3 — MgO	50.25	50.14	49.7	49.6
Yb_2O_3 — MgO	34.9	35.0	65.0	65.0
Yb_2O_3 — MgO	19.2	19.3	80.7	80.7

According to the proposed method of analysis, the content of rare-earth elements is determined by direct complexonometric titration at pH 5 with pyridine and arsenazo as indicator. The solution changes color from violet to orange. In the same solution, direct titration enables us to determine the alkaline-earth element and magnesium, these forming violet complexes with arsenazo [3]. On titrating the solution with Complexone III, the color changes from violet to orange at the point of balance. The proposed method was verified with artificial mixtures (Table 1).

On the basis of the foregoing considerations, the following method was developed for determining the rare-earth and alkaline-earth elements in alloys of the R_2O_3 (SrO, BaO, CaO, MgO) type. A sample 0.2 g in weight is first roasted to constant weight at 1000°C and then dissolved in 4 ml of HCl (1:1). For samples containing rare-earth elements and magnesium or calcium, dissolution may be carried out in sulfuric acid (1:4) with heating (if the sample is not dissolved this is continued until sulfuric acid vapor appears). The solution is cooled, diluted with water, and transferred to a 100-ml measuring flask. An aliquot portion of solution (25 ml) is transferred to a conical flask, and five or six drops of a 0.5% solution of arsenazo are added with 15-20 drops of pyridine; the solution is neutralized with ammonium to pH 5-6 using universal indicator paper, and titrated with a 0.025 M solution of Complexone III until the violet color of the solution turns orange. The content of the alkaline-earth element is determined in the same aliquot portion of solution. For this purpose ammonia is added to the titrated part of the solution until the pH becomes equal to 10. The coloring then reverts to its original tint; titration proceeds with the same solution of Complexone III until the violet turns orange. The titer of the solution of Complexone III is established by reference to standard solutions of the elements under consideration, subject to the same conditions of analysis.

A number of samples were analyzed by the proposed method. For comparison purposes, the rare-earth elements, yttrium, scandium, and the alkaline-earth elements were determined in the same samples by the classical gravimetric method: ytterbium and yttrium by the oxalic acid procedure, and barium, calcium, and strontium by the sulfuric acid technique (Table 2).

It follows from the results obtained that the method proposed for the analysis of such alloys is entirely satisfactory both in accuracy and in the reproducibility of the results.

In analyzing alloys of the rare-earth elements with chromium oxide the decomposition of the sample presented great difficulty. In order to decompose such alloys we tried sintering

TABLE 3. Gadolinium Oxide in Artificial Mixtures in Presence of Hexavalent Chromium

Taken, mg	Found, mg	Relative error, %
92.77	92.77	0
94.63	94.15	—0.4
58.26	57.52	—1.3
38.96	38.49	—1.0
74.2	74.5	+0.5
30.8	30.6	—0.7
33.39	33.39	0
66.50	66.50	0

TABLE 4. Gadolinium Oxide and Chromium Oxide in Samples

Gadolinium oxide, %		Chromium oxide, %	Gadolinium oxide, %		Chromium oxide, %
gravimetric oxalate method	complexonometric method		gravimetric oxalate method	complexonometric method	
—	22.9	76.6	65.0	65.3	34.7
	23.0	76.8	65.0	65.2	34.65
44.1	44.3	55.6	—	68.8	31.0
44.2	44.3	55.75		68.8	31.2
52.4	52.6	47.1	—	69.8	30.3
52.2	52.5	47.2		69.6	30.1

them with a mixture of soda and magnesium oxide at 1000°C in aluminum oxide crucibles. Such crucibles are stable to high temperatures, and furthermore the aluminum passes into solution to only a slight extent when determining the rare-earth elements and is easily masked by sulfosalicylic acid.

Analytical Procedure

A 0.2 g sample is baked with a fivefold amount of mixed sodium carbonate and magnesium oxide (5:1) in aluminum oxide crucibles at 900-1000°C for 2-3 h. The cake is leached with water in a beaker, and with careful heating drops of sulfuric acid (1:4) are added until the cake is completely dissolved. The solution is transferred to a measuring flask after having oxidized the chromium with ammonium persulfate.

Chromium and the rare-earth elements are determined from aliquot portions, the latter by titration with complexone III at pH 5, using an indicator of xylenol orange, after previously combining the aluminum with sulfosalicylic acid. The titer of the solution of Complexone III is established by reference to a standard solution of rare-earth elements. The chromium is determined by the silver persulfate method. The method was verified for artificial mixtures (Tables 3 and 4).

Conclusions

1. We have developed a method of analyzing alloys consisting of the oxides of rare-earth and alkaline-earth elements based on the complexonometric determination of the rare-earth and alkaline-earth elements in one and the same solution at different pH values, using the same indicator.

2. We have proposed a method of analyzing alloys consisting of the oxides of rare-earth elements and chromium oxide. For decomposing these alloys the samples are baked with a mixture of sodium carbonate and magnesium oxide (1:5) in aluminum oxide crucibles at 900-1000°C.

Literature Cited

1. T. N. Nazarchuk and O. I. Popova, Complexonometric Analysis of Metalloceramic and Ceramic Materials and Certain Alloys, Metallurgizdat, Moscow (1965), pp. 23-32.
2. S. P. Onosova, Zavod. Lab., 28:271 (1962).
3. L. Ya. Polyak, Zavod. Lab., 27:7 (1961).
4. G. Brunischolz and R. Cahen, Helv. Chim. Acta, 39:324 (1956).
5. G. Brunischolz and R. Cahen, Helv. Chim. Acta, 39:2135 (1956).
6. H. Flascka, Microchim. Acta, 1:55 (1955).
7. J. C. Fritz et al., Z. analyt. Chem., 30:1111 (1958).
8. Patrowski, Coll. Czech. Chem. Comm., 24:3305 (1959).
9. R. Pribil and V. Vesely, Talanta, 10:899 (1963).

SEPARATION OF ALLOYED CHROMIUM CARBIDES OF THE Cr$_2$C (METASTABLE) AND Cr$_{23}$C$_6$ OR Cr$_7$C$_3$ TYPES ISOLATED FROM STEELS AND ALLOYS

L. V. Zaslavskaya, N. V. Ivanova, and N. F. Lashko

Moscow

The presence of alloying elements, such as Cr, Mo, W, V, N, and others, in steels and alloys under certain conditions promotes the formation of carbide or nitride phases with a widely varying composition, of the Me$_2$C or Me$_2$N type, having a hexagonal crystal structure. The temperature range of stability of these phases in steels and alloys depends on the particular form of alloying involved. These phases often occur with other carbide or nitride phases, which impedes the determination of their chemical composition and any analysis of the distribution of the alloying elements between the various phases.

A particular circumstance only established quite recently is the fact that in a number of extensively used heat-resistant steels containing 12% Cr and alloyed with Mo, W, and other elements the hardening process depends on the specific characteristics of precipitating phases of the Me$_2$C type, their composition, and their degree of dispersion [2, 3, 4]. The phase in question has been assigned the notation Me$_2$X. The Me$_2$X phase may be either a chromium carbide of the Me$_2$C type or a carbonitride Me(C, N). The metastable phase based on chromium carbide Cr$_2$C becomes stable on alloying with Mo, W, V, or N.

N. F. Lashko [1] observed an Me$_2$C phase in an alloy containing 0.2% C, 18% Cr, 8.9% W, 58% Ni with a vanadium content varying from 1 to 4%. In this alloy a phase based on the carbide Cr$_2$C enriched with vanadium (up to 20%) and tungsten (up to 40%) is formed.

The chemical composition of the Cr-base phase of the Me$_2$C type isolated from steels containing 12% Cr has never been completely established, since the Me$_2$C phase is isolated electrolytically together with the chromium carbides Me$_{23}$C$_6$ and Me$_7$C$_3$.

TABLE 1. Chemical Composition of the Steels, %

No. of steel	C	Mn	Si	Cr	Ni	W	Mo	V	Ti	B	Quench temp., °C
1	0.15	0.50	0.50	13.60	3.05	1.66	None	0.20	0.043	0.005	1050 (oil)
2	0.24	0.30	0.46	12.70	1.70	1.77	1.70	0.24	None	None	1050 (oil)
3	0.15	0.38	0.20	14.45	4.47	None	2.48	Nitrogen	~0.08%		1050 (water) cold treatment [-70° for 2h]

TABLE 2. Results of the Differential Chemical Analysis
of the Anodic Deposits Isolated from Steels 1 and 2

No. of steel	Temperature, °C	Time, h	Me_2X						$Me_{23}C_6$						Phase composition
			Amount of phase	Cr	Fe	W	Mo	V	Amount of phase	Cr	Fe	W	Mo	V	
1	540	3	1.60	74.4	10.6	11.2	—	3.7	—	—	—	—	—	—	Me_2X
	560	3	0.94	73.4	12.7	11.7	—	2.1	0.54	64.8	18.5	9.3	—	7.4	Me_2X, $Me_{23}C_6$
	600	3	0.31	67.7	16.1	9.6	—	6.4	1.81	64.6	21.5	12.1	—	1.7	Me_2X, $Me_{23}C_6$
	680 680	4 4	1.97	60.0	11.6	10.1	14.7	3.0	1.65	58.4	22.3	7.2	9.0	3.0	Me_2X, $Me_{23}C_6$
2	400 680	100 4	2.47	64.0	10.9	9.3	11.9	2.6	0.65	56.7	30.0	4.6	7.6	1.0	Me_2X, $Me_{23}C_6$
	450 680	100 4	2.89	62.0	13.8	9.0	11.0	2.0	0.52	53.6	33.1	6.7	7.6	1.3	Me_2X, $Me_{23}C_6$
	550 680	100 4	2.51	63.8	13.9	9.2	11.1	2.0	1.57	58.8	23.1	5.4	10.1	1.25	Me_2X, $Me_{23}C_6$

In chemical composition the Cr-base Me_2C phase is, on the one hand, analogous to the carbides Mo_2C, W_2C and the nitrides Cr_2N, Nb_2N with a hexagonal structure, since all these break down under the action of hydrogen peroxide, and, on the other hand, the Cr-base phase is decomposed by acids in the same way as the cubic chromium carbide, which remains immune to the action of hydrogen peroxide.

The instability of the Me_2C phase with respect to hydrogen peroxide was used in order to separate it from the cubic chromium carbide. We took three types of steel with the compositions indicated in Table 1. For the electrolytic isolation of the phases we used a 5% solution of hydrochloric acid in hydrolytic ethyl alcohol with a current density of 0.01-0.02 A/cm². X-ray structural analysis of the anodic deposits isolated from steels 1 and 2 revealed chromium carbides of the $Me_{23}C_6$ and Me_2C types, the latter having a hexagonal structure with lattice constants $a = 2.86°$, $c = 4.47$ Å. The phases isolated from steel 3 consisted of chromium carbonitride $Me_2(CN)$ and $Me_{23}C_6$.

The anodic deposits were treated with alkali in order to remove contaminants of the tungstic and molybdic acid types, passing into the anodic deposit in the course of electrolysis. The method of separating the phases reduces to the following: the anodic residue, washed free from alkali, is placed in a 0.5-liter beaker, 100 ml of hydrogen peroxide are added, and the result is heated on a hot-plate (thick asbestos) for 2h, hydrogen peroxide being added as decomposition proceeds.

Under the action of the hydrogen peroxide the cubic or trigonal chromium carbide is passivated and remains in the insoluble residue. The Me_2C or $Me_2(C, N)$ phase is decomposed and passes into the filtrate. Chemical analysis of the filtrate and insoluble residue indicates the concentrations of the elements in the Me_2C phase and the $Me_{23}C_6$ or Me_7C_3 carbides. The results of a differential chemical analysis are presented in Table 2.

Figure 1 shows the content of the elements in the separated phases for steel 2. We see that in the Me_2C phase the Cr, Mo, and W content is higher than in the carbide $Me_{23}C_6$, while the Fe content is much smaller.

Figure 2 shows the change in the concentration of Me_2C and $Me_{23}C_6$ isolated from steel 1 in relation to the form of heat treatment applied. After quenching from 1050 and tempering at 540°C for 3 h, only the Cr-base carbide Me_2C separates to the extent of 1.60%. On raising the tempering temperature to 560°C for the same period, the amount of Me_2C diminishes and the cubic chromium carbide appears to the extent of 0.60%; on further raising the tempering temperature to 600°C the amount of Me_2C falls still further, while the amount of the cubic carbide

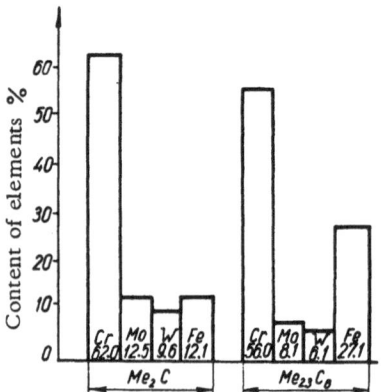

Fig. 1. Composition of the carbides Me_2C and $Me_{23}C_6$ for steel 2.

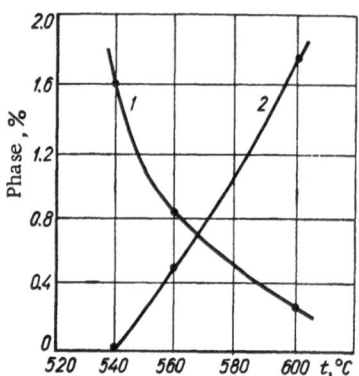

Fig. 2. Concentration of carbides Me_2C (1) and $Me_{23}C_6$ (2) in steel 1 as a function of tempering temperature (heating for 3 h).

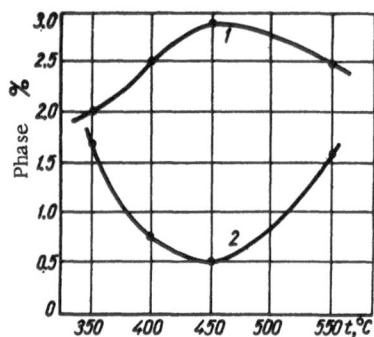

Fig. 3. Effect of the temperature of supplementary heating on the kinetics of Me_2C (1) and $Me_{23}C_6$ (2) for steel 2.

increases by a factor of 3.5. This behavior of the Me_2C may be explained by its relative stability at low temperature and its metastability at higher tempering temperatures. The cubic chromium carbide constitutes the stable phase in the latter case.

Figure 3 shows the kinetic characteristics of the formation of Me_2C and $Me_{23}C_6$ in relation to the temperature of supplementary heating for steel 2. After quenching and tempering at 680°C for 4 h a considerable quantity of chromium carbide of the Me_2C type (1.97%) and the carbide $Me_{23}C_6$ (1.66%) is isolated.

After additional heating at 400 or 450°C for 100 h the amount of cubic carbide diminishes and the amount of Me_2C correspondingly increases. On raising the temperature of supplementary heating to 550°C for the same period the opposite effect occurs; the cubic carbide becomes more abundant and the Me_2C diminishes. Thus as in steel 1 the Me_2C is relatively stable at the lower temperatures (400-450°C) and $Me_{23}C_6$ at the higher (550-650°C) so that a redistribution of the alloying elements takes place between the phases.

In view of the fact that nitrogen had been incorporated in steel 3, the proportion of this element in the anodic residues had to be determined. The introduction of nitrogen into the steel makes the $Me_2(C, N)$ phase more stable in view of the formation of chromium carbonitride, which is stabler than the chromium carbide of type Me_2C.

Table 3 shows the results of a differential chemical analysis of the anodic deposits isolated from steel 3. The amount of carbon is given by calculation. We see from the table that, in the course of tempering at 500°C for 1 h, a small quantity of chromium carbonitride of the

TABLE 3. Results of the Differential Chemical Analysis
of the Anodic Deposits Isolated from Steel 3

Tempering temperature, °C *	Amount of phase	Cr	Fe	Mo	N	C	Amount of phase	Cr	Fe	Mo	Ni	C	Phase composition
500	0.22	72.0	9.0	9.0	2.3	7.6	—	—	—	—	—	—	Me_2X (C, N)
450	0.40	75.0	7.5	7.5	2.5	7.5	—	—	—	—	—	—	Me_2X (C, N)
520													
650	1.74	65.5	13.8	10.9	2.3	7.4	0,88	55.7	29.5	6.8	2.2	5,1	Me_2X, $Me_{23}C_6$

* Tempering time 1 h.

$Me_2(C, N)$ type is formed; after double tempering at 450 and 520°C for the same time the amount of chromium carbonitride increases. At 650°C the $Me_2(C, N)$ content rises to 1.74% and 0.90% of the cubic carbide appears.

It should be noted that the temperature range of stability of the Me_2C phase in steel 1 (alloyed with tungsten only) is narrower than in the other two steels. After tempering at 560°C the amount of Me_2C starts falling; at 600°C it is only 0.31%, while the cubic chromium carbide reaches 1.81%.

The Me_2C phase in steel 2 (alloyed with molybdenum and vanadium as well) is more stable. At 550°C it appears to the extent of 2.51% together with 1.57% of the cubic chromium carbide.

In steel 3 (alloyed with molybdenum and nitrogen) the $Me_2(C, N)$ phase is still stabler. At 650°C there is more of this phase than of the cubic carbide (1.74 and 0.88% respectively).

Conclusions

1. We have proposed a quantitative method of separating carbides or carbonitrides of the $Cr_2(C, N)$ type and chromium carbides of the $Me_{23}C_6$ type in their mutual presence.

2. The amount of iron in the Me_2X Cr-base phase is much smaller than in the cubic carbide $Me_{23}C_6$.

3. The temperature range corresponding to the existence of the metastable phase depends on the alloying characteristics of the steel. The phase is stabilized to a greater extent by molybdenum and nitrogen than by tungsten.

Literature Cited

1. N. F. Lashko and E. Ya. Rodina, Fiz. Met. Metallov., Vol. 5, No. 2 (1957).
2. K. W. Andrews et al., J. Iron Steel Inst., 193(3):304 (1959).
3. K. L. Irvin et al., J. Iron Steel Inst., 195(4):386 (1960).
4. K. L. Irvin et al., J. Iron Steel Inst., 200(10):820 (1962).

CHEMICAL AND ELECTROCHEMICAL METHODS OF SEPARATING THE CARBIDES MeC AND THE CARBONITRIDES Me(C,N) OF GROUP IV AND V METALS

G. G. Georgieva, N. F. Lashko, and K. P. Sorokina

Moscow

The interaction of the carbides MeC, nitrides MeN, and carbonitrides Me(C, N) of Group IV and V metals with solutions of mineral acids with or without the superposition of a current differs from the interaction of the carbides and nitrides of Group VI-VIII metals. This fact enables us to isolate individual carbides MeC or nitrides MeN from mixtures of carbides or nitrides, or from steels and alloys containing these phases. It is a more complicated matter to separate the carbides MeC and nitrides MeN of Groups IV and V from each other.

The synthetic carbides MeC of Group IV and V metals differ as regards their interaction with mineral acids from the carbides occurring in steels and alloys melted in open furnaces, since the latter also contain nitrogen (arising from the charge materials and the air) and as a rule alloying elements (Cr, Mo, W, etc.) also present in the steels and alloys.

We have found that in aqueous and alcohol electrolytes containing Cl', SO_4'', PO_4'', and NO_3' ions carbides and nitrides of the TiC and TiN type are passivated and isolated quantitatively from various steels and alloys, in contrast to the carbides $Me_{23}C_6$ and Me_7C_3, which are only passivated and isolated quantitatively in electrolytes containing Cl'.

Nickel-base solid solutions in steels and alloys exist in an active state and dissolve in electrolytes containing Cl', NO_3', SO_4'', and PO_4''' ions at potentials at which the carbide TiC and the hardening phase $Ni_3(Al, Ti)$ are in the passive state; this enables us to isolate the $Ni_3(Al, Ti)$ phase quantitatively together with the carbide TiC and the carbonitride Ti(C, N).

The separation of $Ni_3(Al, Ti)$ from TiC or Ti(C, N) may be achieved by using the capacity of $Ni_3(Al, Ti)$ to dissolve on boiling in dilute aqueous solutions of hydrochloric and sulfuric acids. The TiC, Ti(C, N), and TiN fail to dissolve in these.

It is of practical interest to discover the conditions for the chemical separation of $Ni_3(Al, Ti)$ from MeC or Me(C, N). The carbide and carbonitride phases MeC and Me(C, N) encountered most frequently in practical steels and alloys involve three elements: Ti, Nb, and V.

The results of our own experiments on the chemical separation of MeC carbides in mineral acids on boiling are shown in Table 1 together with published data [1, 2]. As regards solu-

TABLE 1. Solubility of the Carbides MeC of Group IV
and V Metals in Strong Acids and Acid Mixtures
and in the Presence of Oxidizing Agents (on Boiling)

Type of carbide	HCl	H_2SO_4	HNO_3	$HCl + HNO_3$	$H_2SO_4 + HNO_3$	$HCl + H_2O_2$	$H_2SO_4 + H_2O_2$	HF	$HF + HNO_3$
TiC	n.	n.	sl.d	d.	d.	d.	d.	–	d.
ZrC	n.	n.	sl.d	d.	d.	d.	d.	–	d.
HfC	n.	n.	sl.d	d.	d.	–	–	–	d.
VC	n.	n.	d.	d.	d.	d.	d.	n.	d.
NbC	n.	n.	n.	sl.d	n.	d.	d.	n.	d.
TaC	n.	n.	n.	d.	d.	–	–	–	d.

Note. According to Blanter [2] NbC dissolves in H_2SO_4; here and in
Table 2 n denotes "not dissolved," d denotes "dissolved," and sl.d.
denotes "slightly dissolved."

TABLE 2. Solubility of Group IV and V Metals in
Mineral Acids and Acid Mixtures*

Metal	HCl	H_2SO_4	HNO_3	$HNO_3 + HCl$	HF	$HF + HNO_3$
Ti	d.	d.	d.	d.	d.	d.
Zr	n.	n.	n.	d.	d.	–
Hf	n.	n.	n.	d.	d.	–
V	n.	n.	d.	d.	d.	d.
Nb	n.	n.	n.	n.	n.	d.
Ta	n.	n.	n.	n.	n.	d.

* B. V. Nekrasov, Course of General Chemistry [in Russian],
Goskhimizdat, Moscow (1953).

bility in mineral acids, the MeC carbides are more stable than the metals which they incorporate (Table 2).

The greatest solubility occurs for the carbides of the Group IV and V metals of the fourth period (Ti and V); the solubility diminishes for combined V and VI period metals. It is well known that the introduction of oxidizing agents into the solution (particularly HNO_3 and H_2O_2) favors the dissolution of MeC carbides.

TABLE 3. Chemical Composition of the
Alloys Studied (Nitrogen Content 0.03%)*

C	Cr	Ti	Al	Mo	Nb
0.024	14.36	2.62	1.47	2.99	1.96
0.053	14.36	2.62	1.47	2.99	1.96
0.087	14.36	2.62	1.47	2.99	1.96
0.12	14.33	2.7	1.65	2.93	2.14
0.20	14.33	2.7	1.65	2.93	2.14
0.31	14.33	2.7	1.65	2.93	2.14

*Phase composition α', (Nb, Ti) (C, N), TiN, Me_3B_2

Judging from the results of Table 1, the chemical separation of NbC and VC should present little difficulty. In studying the phase composition of a number of austenitic steels containing niobium and vanadium (particularly steel ÉI481) it was found that the carbonitride Nb(C, N) and the carbide VC formed in them separated on boiling in nitric acid (d = 1.4 g/cm³) [4]. The vanadium carbide dissolved.

We made a special study of the separation of the phases Ni_3(Al, Ti), Nb(C, N), TiN formed in nickel alloys of the ÉI-698 type. The chemical composition of the alloys is shown in Table 3. Heat treatment was as follows: quenching from 1120°C (8 h) + 1000°C (4 h) + 775°C (16 h) + 650°C (16 h); cooling after each operation in air.

According to x-ray structural analysis,* the alloy ÉI-698 constitutes a many-phased system including an austenite solid solution together with the phases Ni_3(Ti, Al), (Nb, Ti), (C, N), TiN and the boride phase $(Mo, Cr)_3B_2$.

The phases may be isolated and determined separately in the following manner:

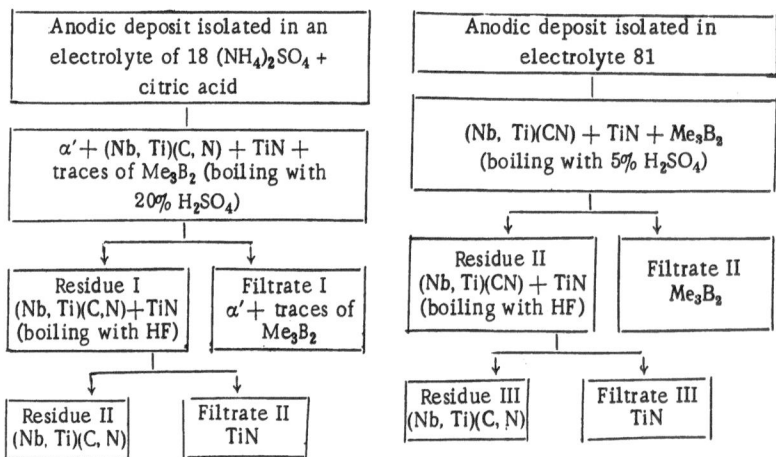

The anodic deposit isolated in the aqueous electrolyte (18) containing 10 g of ammonium sulfate and 30 g of citric acid (quite enough to prevent the precipitation of niobic acid from solid solution) is boiled with 20% H_2SO_4, by virtue of which the α' phase and the boride Me_3B_2 pass into solution. The content of boride phase may be neglected, since the boron in the alloy amounts to no more than 0.01%, and the boride phase is usually not isolated completely in aqueous electrolytes containing sulfates.

The anodic deposit isolated in the alcohol electrolyte 81 (which contains the Cl' ion), consisting of (Nb, Ti) (C, N), TiN, and Me_3B_2 (the latter is completely isolated in this electrolyte), are boiled with 5% sulfuric acid for 1.5 h, during which the Me_3B_2 is dissolved completely.

On boiling the anodic deposit isolated from the ÉI-698 alloy, containing niobium carbonitride and a small quantity of titanium nitride, we observed the selective dissolution of the titanium (Table 4). In the filtrate we found a large amount of titanium after boiling with HNO_3. According to x-ray structural analysis, before boiling with HNO_3 the unit-cell parameter of the carbonitride (Nb, Ti) (C, N) equaled 4.393 Å but after the selective dissolution of the titanium it changed to 4.397 Å.

In a mixture of hydrochloric and nitric (or sulfuric and nitric) acids and in the presence of hydrogen peroxide, the anodic deposit containing the phases (Nb, Ti) (C, N) and TiN dissolves completely.

By virtue of the chemical separation of the phases we determined the composition of the α' phase. The content of α' phase may be determined by two methods independently of the amount of carbon in the alloy. The carbide and carbonitride phases of niobium and titanium are quantitatively isolated in an alcohol electrolyte containing HCl and also in aqueous electrolytes containing ammonium sulfate.

---

*The x-ray investigation was conducted by K. V. Smirnova.

TABLE 4. Solubility of Isolated Carbonitrides in Acids and Acid Mixtures
(Alloy of the ÉI-698 Type, C = 0.31%)

Phase	Solution	Time	Residue		Filtrate	
			Nb	Ti	Nb	Ti
(Nb, Ti) (C, N) + TiN	HF (conc.)	1.5 h	1.20	0.49	*	0.03
	HF (50%)	1.5 h	1.26	0.48	*	0.04
	HNO$_3$ (conc.)	3 h	1.16	0.07	*	0.45
	HNO$_3$ (conc.)	5 h	1.18	0.08	*	0.44
	50 ml H$_2$SO$_4$ (1:4) + 2 ml HNO$_3$	5 min	*	*	1.20	0.55
	50 ml HCl + 2 ml HNO$_3$	5 min	*	*	1.24	0.53
	HNO$_3$ + 1% NaF	5 min	*	*	1.22	0.54

*Not found

TABLE 5. Proportions of Elements in the
Ni$_3$(Ti, Al) Phase, %

Carbon in alloy, %	Ni	Ti	Cr	Mo	Nb	Al	Sum of Elements %
0.024	14.72	1.50	0.49	0.12	0.60	0.95	18.90
0.053	14.48	1.49	0.49	0.12	0.60	0.93	18.58
0.087	13.00	1.23	0.50	0.11	0.50	0.84	16.29
0.12	12.31	1.16	0.46	0.10	0.48	0.83	15.47
0.20	11.16	1.03	0.44	0.10	0.45	0.70	13.88
0.31	8.62	0.92	0.40	0.08	0.41	0.62	11.24

TABLE 6. Proportions of the Elements in the
Phases (Nb, Ti) (C, N) + TiN + Me$_3$B$_2$, %

Carbon in alloy, %	Ni	Nb	Ti	Cr	Mo	Sum of Elements, %
0.024	<0.01	0.24	0.13	0.02	0.06	0.45
0.053	<0.01	0.25	0.14	0.03	0.09	0.51
0.087	<0.01	0.33	0.18	0.04	0.10	0.65
0.120	<0.01	0.52	0.22	0.04	0.11	0.89
0.20	<0.01	0.73	0.33	0.04	0.16	1.26
0.31	0.01	1.24	0.54	0.08	0.26	2.12

The amount of α' phase determined by the chemical phase separation of the anodic deposits agrees with that derived from the difference between the results based on two methods of electrochemical isolation in electrolytes containing Cl' or SO$_4''$ ions (within the limits of experimental error). The average α' phase content derived from the different methods is shown in Table 5 as a function of carbon content. The amount of α' phase falls from 18.9 to 11.2% as the carbon content rises from 0.02 to 0.31%. The carbonitride and nitride phases meanwhile increase from 0.45 to 2.12% (Table 6). The change in the α'-phase content and the proportion of carbides with carbon content in the alloys is illustrated in Fig. 1.

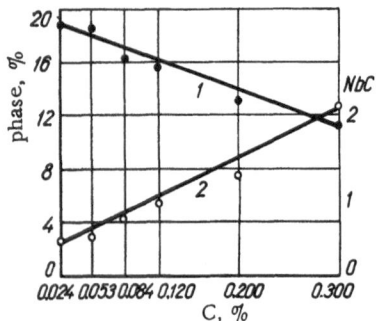

Fig. 1. Amount of α' phase (1) and NbC (2) in alloy ÉI-698 in relation to the carbon content.

Conclusions

1. Carbides and carbonitrides of the MeC and Me(C, N) types derived from Group IV and V metals are isolated quantitatively in electrolytes containing Cl', NO_3', SO_4'', and PO_4''' ions; hence the composition of the Ni_3Al-base hardening phase in heat-resistant alloys containing MeC, or Me(C, N) in addition to the latter, may be determined by two methods: by the chemical separation of the anodic deposit isolated in an aqueous electrolyte containing NO_3', SO_4'', or PO_4''' ions, and from the difference between the results derived from the double electrolytic isolation of the phases in aqueous and alcohol electrolytes (the latter containing the Cl' ion).

2. On prolonged boiling of the carbonitride (Nb, Ti)(C, N) with HNO_3 (3-6 h) the titanium dissolves selectively.

Literature Cited

1.   Analysis of Refractory Compounds [in Russian], Metallurgizdat (1962), p. 50.
2.   M. E. Blanter, Method of Studying Metals and Analyzing Experimental Data [Russian translation], Metallurgizdat (1952), p. 330.
3.   N. I. Blok, N. F. Lashko, and K. P. Sorokina, Phase Composition, Structure and Properties of Alloy Steels and Alloys [in Russian], Mashinostroenie, Moscow (1965), p. 28.
4.   K. P. Sorokina, Zavod. Lab., No. 3, p. 253 (1961).

ELECTROLYTIC ISOLATION OF THE σ PHASE FROM HEAT-RESISTANT STEELS AND THE DETERMINATION OF ITS COMPOSITION

E. F. Yakovleva, I. M. Dubrovina, and L. V. Stegnukhina

Central Scientific-Research Institute of Ferrous Metallurgy

The conditions governing the formation of the σ phase and its effect on the properties of steels and alloys have never finally been elucidated. It is therefore extremely important to conduct further investigations into the development of methods of differential electrochemical analysis for steels and alloys containing the σ phase and also other phase constituents. We studied austenitic boiler steel ÉP17 and a heat-resistant steel (ÉP508) intended for prolonged service at high temperatures with the following chemical composition in %:

	C	Si	Mn	Cr	Ni	Ti	W	Mo	Nb	B
ÉP17	0.08	0.30	1.72	19.2	14.3	—	2.57	—	0.93	0.005
ÉP508	0.08	0.50	0.5	13.0	18.0	1.3	1.3	1.3	—	0.005

According to a preliminary microstructural analysis, the principal phases in ÉP17 were $Nb(C, N)$, Fe_2W, $M_{23}C_6$, and in ÉP508 Ni_3Ti, $Fe_2(W, Mo)$, $Ti(C, N)$. In a sample aged at 700°C for 5000 h (ÉP17) and in a sample of ÉP508 aged at 700°C for 5688 h we noted the appearance of the σ phase. Parts made from these steels are employed under long-service conditions. An analy-

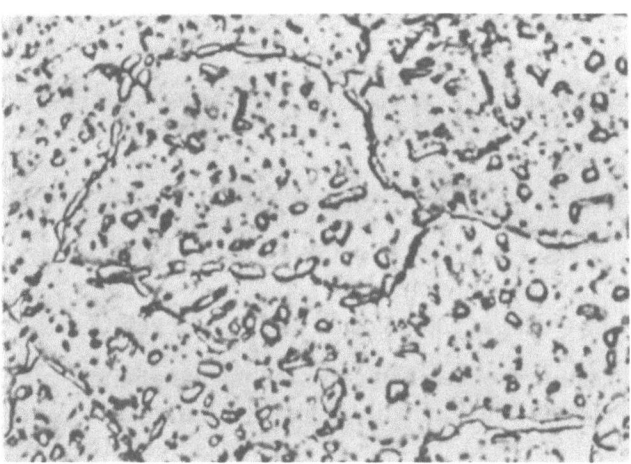

Fig. 1. Microstructure of EP17 steel with large σ-phase precipitates.

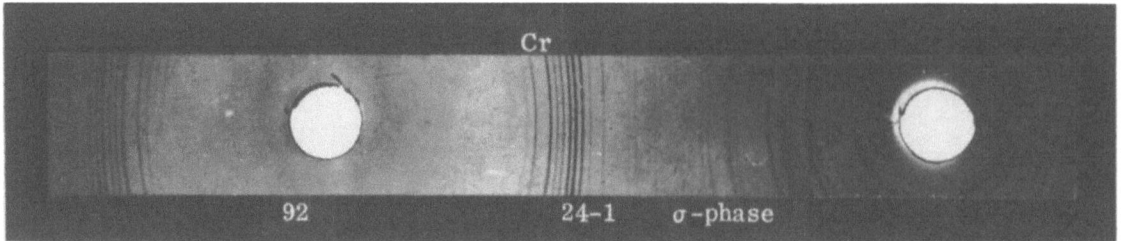

Fig. 2. X-ray diffraction pattern of the pure σ phase.

sis of the phase transformations associated with various forms of heat treatment is accordingly of particular interest. It is well known from published data that the hardness of tensile strength of steel rise and the ductility falls when the σ phase develops. In rare cases the σ phase may constitute a hardening agent. The precipitation of the σ phase along grain boundaries is nevertheless undesirable, since it tends to deprive them of chromium. This leads to a tendency for intercrystallite corrosion and embrittlement of the material.

Various stoichiometrical compositions have been assigned to the σ phase in the literature; these include $CrMn_3$, CrCo, and FeCr. The iron in the σ phase may be replaced by Ni, Co, or Mn and the chromium by V, Mo, or W; no Nb or Ta have yet been found in the σ phase. We ourselves studied a σ phase of the FeCr type. Figure 1 shows the microstructure of ÉP17 steel with large σ phase precipitates.

N. F. Lashko indicates that strong carbide-forming elements (Nb, Ti, Zr) promote the formation of the σ phase. In ÉP17 and ÉP508 steels these elements do in fact occur.

For the electrolytic separation of the σ phase from ÉP17 steel we tested a number of electrolytes:

Electrolyte Composition	Electrolytic Conditions
1150 ml methyl alcohol and 50 ml HCl	Current density 0.05 A/cm², temperature 5°C, duration 1-1.5 h
220 g KCl, 200 ml HCl, 50 g ammonium citrate in 1 liter of water	Current density 0.05 A/cm², temperature 20°C, duration 1-1.5 h
5 g of oxalic acid, 200 ml HCl in 1 liter of water	Current density 0.05 A/cm², temperature 20°C, duration 1-1.5 h

X-ray structural analysis* confirmed the isolation of Nb(C, N), $M_{23}C_6$, and Fe_2W in all the electrolytes; only in a sample aged at 700°C for 5000 h was the $M_{23}C_6$ phase replaced by a phase of the FeCr type with a tetragonal crystal lattice and the parameters a = 8.79 kxu, c = 4.55 kxu.

In order to confirm the presence of the σ phase we carried out a differential chemical analysis based on the treatment of the anodic deposit with a mixture of equal volumes of a 20% solution of tartaric acid and 30% H_2O_2. The σ phase is passivated and remains in the deposit, while the Nb(C, N) and Fe_2W dissolve completely. The completeness of the phase separation is confirmed by x-ray structural analysis. We obtained an x-ray diffraction pattern of the pure σ phase to serve as a standard in solving other x-ray patterns (Fig. 2).

---

*The x-ray structural analysis was conducted by V. A. Belyaeva.

Chemical analysis of the deposit was carried out after its electrolytic separation in an electrolyte of the same composition as that used for the x-ray structural analysis. However, the chemical analysis indicated that the σ phase was not always precipitated quantitatively in this electrolyte, since parallel analyses yielded poor reproducibility.

The σ phase was isolated in the most quantitative manner in an electrolyte containing hydrochloric and oxalic acids. In this electrolyte we separated 3.89 wt.% of the σ phase and in the previous one 2 wt.%.

In sample 24-11, tempered (aged) at 700°C for 5000 h, the proportions of the elements in the phase were

Element	Wt.%	At.%
Cr	43.96	48.22
W	7.71	2.35
Fe	48.32	49.33

In this case the atomic ratio $Fe/(Cr+W) = 0.97$, i.e., according to the chemical analysis it approaches a stoichiometric ratio of unity. In addition to the σ phase, the $Nb(C,N)$ and an Fe_2W-base Laves phase were determined in this alloy. The composition of the $Nb(C,N)$ phase was determined after treating the anodic deposit with HF, which fails to dissolve $Nb(C,N)$. The composition of the Laves phase was calculated from the difference between the chemical analysis of the whole anodic deposit and the sum of the niobium carbonitride and σ phase. The results of the differential chemical analysis are presented in Table 1. The elements were determined as follows: niobium photocolorimetrically with cyanoformazon, tungsten photocolorimetrically with thiocyanate, iron photocolorimetrically with sulfosalicylic acid, and chromium by the volumetric silver persulfate method.

In order to isolate the σ phase from ÉP508 heat-resistant steel after aging at 700° for 5688 h we used an electrolyte containing 20% of potassium chloride and hydrochloric acid and 8% of ammonium citrate. The electrolytic conditions were: current density 1 A/cm², electrolyte temperature 0°C, duration of the electrolysis 10-15 min. X-ray structural analysis of the anodic deposit showed the σ phase together with a Laves phase based on $Fe_2(W,Mo)$ and $Ti(C,N)$. In order to separate the σ phase chemically from the $Fe_2(W,Mo) + Ti(C,N)$, we used the method developed for ÉP17 steel, i.e., treatment of the anodic deposit with a mixture of tartaric acid

TABLE 1. Results of A differential Chemical
Analysis of the Phases Isolated from ÉP17 Steel
(Heat Treatment at 700°C for 5000 h)

Phase	Conc. of elements in the phases		Concentration of phase	Atomic ratio
	wt.%	at.%		
(Fe, Cr)$_2$ W	—	—	0.94	$\dfrac{Fe+Cr}{W} = 2.09$
Fe	0.13	23.50	—	—
Cr	0.23	44.66	—	—
W	0.58	31.83	—	—
Nb (C, N)	—	—	—	—
Nb	0.74	—	—	—
W	0.014	—	—	—
N$_2$	0.017	—	—	—
σ phase	—	—	3.89	$\dfrac{Fe}{Cr+W} = 0.97$
Cr	1.71	48.30	—	—
W	0.30	2.35	—	—
Fe	1.88	49.33	—	—

TABLE 2. Results of a Differential Analysis
of the Phases Isolated from ÉP508 Steel
(Sample 71-1)

Phase	Conc. of elements in the phases		Concentration of phase	Atomic ratio
	wt.%	at.%		
Ti (C,N)	—	—	—	—
Ti	0.30	—	—	—
W	—	—	—	—
Mo	0.03	—	—	—
N$_2$	0.08	—	1.997	$\dfrac{\text{Fe} + \text{Cr}}{\text{W} + \text{Mo}} = 1.9$
Fe$_2$(W, Mo)	—	—	—	—
Fe	0.71	50.25	—	—
W	0.53	11.42	—	—
Cr	0.197	15.09	—	—
Mo	0.56	23.22	—	—
σ phase	—	—	—	$\dfrac{\text{Fe}}{\text{Cr} + \text{W} + \text{Mo}} = 1.2$
Fe	0.21	55.45	—	—
Cr	0.11	31.12	—	—
W	0.08	6.78	—	—
Mo	0.04	6.64	—	—

and perhydrol, in which the Laves phase and titanium carbonitride dissolve, while the σ phase remains in the residue. The completeness of the phase separation was confirmed by x-ray structural analysis conducted by V. A. Belyaeva and V. G. Kostogonov. Differential chemical analysis of the σ phase revealed the presence of Fe, Cr, W, and Mo in the σ phase. The atomic ratio of the iron to the Cr + W + Mo sum equaled 1.2, approaching stoichiometric ratio. The results of the differential chemical analysis are presented in Table 2.

Conclusions

1. We have developed a method for the differential chemical analysis of Nb(C, N), Fe$_2$W, and the σ phase isolated from austenitic steel ÉP17 and Ti(C, N), Fe$_2$(W, Mo), and the σ phase isolated from heat-resistant steel ÉP508.

2. On the basis of the method so developed we have determined the chemical composition of the σ phase in these steels.

3. The proposed method has been introduced into practical laboratory work.

INTERACTION OF REFRACTORY GROUP V METALS WITH ZINC

A. P. Obukhov, V. N. Gurin, I. R. Kozlova, Z. P. Terent'eva, and T. I. Mazina

Physicotechnical Institute, Academy of Sciences of the USSR

In view of the possibility of producing refractory compounds of transition metals [2], particularly V, Nb, and Ta, in a zinc melt, we studied the interactions of these metals with zinc. Existing data relating to the V–Zn, Nb–Zn, and Ta–Zn systems are incomplete and inaccurate. The most extensive and detailed investigations into these systems have been conducted by Shubert and others [7, 9]. Table 1 gives the structural characteristics of intermetallic phases belonging to the systems in question. We see that the structure and parameters of the unit cell have only been determined for a few of these compounds. The Nb–Zn system is the one which has been most studied.

We studied the interaction of refractory metal powders with an excess of molten zinc at the boiling point of 907°C. The boiling time fluctuated between 10 and 30 min. The amount of metal in the compositions studied never exceeded 10 wt.%. Oxidation of the surface of the boil-

TABLE 1. Structural Characteristics of the Phases
in the V–Zn, Nb–Zn, and Ta–Zn Systems

Phase	Structure	Lattice constants, Å			Reference
		a	b	c	
V_4Zn_5	—	8.91	—	3.22	[8]
$VZn_{\sim 3}$	Cu	3.84	—	—	This paper
VZn_3	Cu_3Au	—	—	—	[8]
NbZn	—	—	—	—	[10, 11]
Nb_2Zn_3	—	—	—	—	[2, 11]
$NbZn_2$	Ni_2Mg	5.05	—	16.32	[7]
$NbZn_{2.0-2.7}$	Cu	3.9325	—	—	This paper
$NbZn_3$	Cu_3Au	3.932 3.934	—	—	[9] [6]
Ta_4Zn_5 ?	HF_4Zn_5	—	—	—	[3]
$TaZn_2$?	—	—	—	—	[3]

81

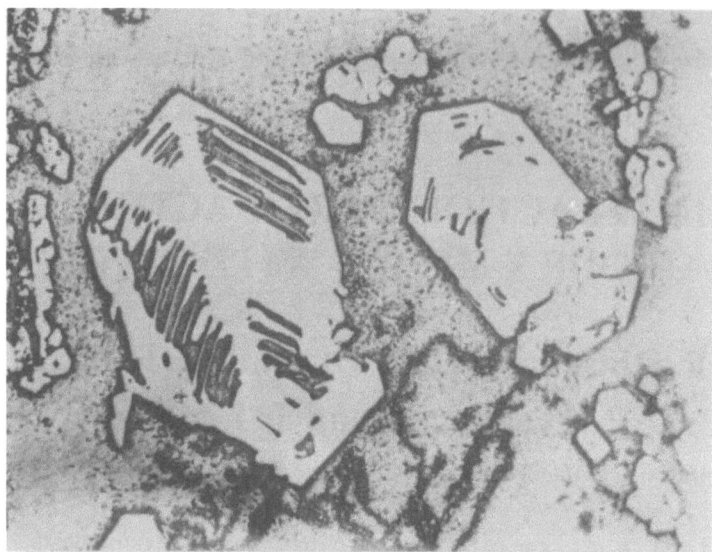

Fig. 1. Microsection containing the NbZn$_{2.0-2.7}$ phase;
magnification 64 times.

ing zinc was prevented by a flow of argon. By collecting a sample in a quartz test tube, cooling, and treating it in (1:1) HCl we were able to separate niobium zincide from the zinc and obtain it in the form of a powder. Vanadium zincide was less stable and dissolved in the acid with the zinc. Tantalum zincide was separated together with tantalum.

Since the niobium zincide was obtained in pure form we were able to examine it in more detail. Microsections indicate the form of the precipitates very clearly; well-faced grains apparently constituting individual crystals are clearly to be seen. In shape these are hexagonal with unequal sides. In one microsection (Fig. 1) dendrites and skeletal growth forms with inclusions of the mother solution are observed. The Vickers microhardness fluctuates from 90 to 338 kg/cm^2 for different grains, indicating a wide range of homogeneity.

Chemical analysis of niobium zincide [1, 3] was carried out as follows: A 0.2-g sample was dissolved by heating in a mixture of concentrated acids (4 ml HF + 2 ml HNO$_3$), 2-3 ml of H$_2$SO$_4$ were added, the result was evaporated until SO$_3$ vapor appeared, and the greater part of the niobium was separated by tartaric acid hydrolysis. In the filtrate the zinc was precipitated in the form of the sulfide, filtered, roasted, weighed as ZnO, and converted to Zn, the conversion factor being 0.8034. After the removal of the hydrogen sulfide, the Nb in the filtrate was completely precipitated by means of Cupferron, roasted, and weighed as Nb$_2$O$_5$. The Nb$_2$O$_5$ deposits (after their separation by tartaric acid hydrolysis and Cupferron respectively) were added together and converted to Nb, the conversion factor being 0.6990. The method was verified on artificial mixtures. The mean square error was ± 0.5 for Nb and ± 0.4 for Zn. The results of the chemical analysis showed that the composition of the compound agreed with the averaged formula NbZn$_{2.35}$. The Nb:Zn ratio fluctuated from NbZn$_2$ to NbZn$_{2.7}$, indicating a wide range of homogeneity of this intermetallic compound.

X-ray structural analysis of the powder suggested the structure of copper for niobium zincide; this was supported by a calculation of the reflection intensities on the Debye photograph based on the assumption that the Nb and Zn atoms were arranged in a disordered manner in the lattice points of a copper lattice. The lattice constant was 3.9325 Å, the accuracy of its determination being ± 0.0001 Å; the interatomic spacing was 2.78 Å, quite close to the sum of the atomic radii of Nb and Zn (2.84 Å). It is interesting to compare our own data regarding the

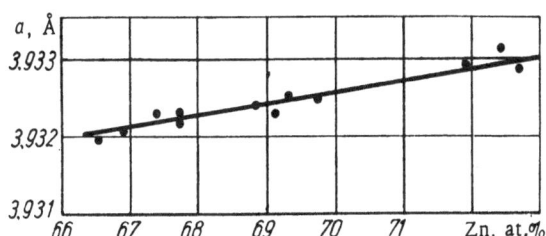

Fig. 2. Relationship between the unit-cell parameters (lattice constants) and zinc content of $NbZn_{2.0-2.7}$.

structure of $Nb-Zn_{2.0-2.7}$ with Vold's data [10] relating to the compound $NbZn_3$; Fold assigned this compound a structure of the Cu_3Au type with a lattice constant of 3.934Å. It is easy to see the relation between the structures of Vold's compound and ours.

The structure of $NbZn_3$ is ordered on the basis of the copper structure. The zinc occupies fixed positions in the center of the faces. Vold looked for disordering at 470-870°C, but found none. Our own results show that disordering occurs when the composition deviates from stoichiometrical.

We studied the relationship between the lattice constant and the composition. A change in zinc concentration by 6.2 at.% corresponded to a lattice-constant fluctuation of 0.0009Å. There was a linear relationship between the lattice constant and zinc content (Fig. 2). This relationship may be understood by comparing the ionic radii of Nb and Zn rather than the atomic radii, as is usually done for intermetallic compounds.

We thus conclude that the metallic bond in the compound is by no means pure; it contains an appreciable trace of a stronger, covalent bond, and this is confirmed by the properties of the compound, including its stability. The reduction in the interatomic distance relative to the sum of the atomic radii of Nb and Zn indicates the stronger interaction.

The solubility of the intermetallic compound was studied in weak and concentrated alkalis and in concentrated acids, both in the cold and on heating. We see from Table 2 that the compound is almost insoluble in boiling concentrated HNO_3; it dissolves rather better in HCl and alkalis, and completely in boiling concentrated H_2SO_4. Analysis of the filtrate revealed a disruption in stoichiometry; on dissolving in acids the Zn:Nb ratio increases, and on dissolving in alkalis (particular weak alkalis) it diminishes. Thus the chemical individuality of each component of the compounds makes itself apparent.

TABLE 2. Behavior of $NbZn_{2.0-2.7}$ in Various
Media in the Cold (One Day) and on
Heating (2 h)

Medium	In the cold			On heating		
	Insoluble residue, %	Amount in solution, %		Insoluble residue, %	Amount in solution, %	
		Nb	Zn		Nb	Zn
HCl	19.23	27.20	53.47	14.49	30.45	50.66
HNO_3	96.96	—	—	99.20	—	—
H_2SO_4	69.18	9.51	19.02	None	Original	
$3HCl + HNO_3$	64.82	10.43	22.77	36.22	19.47	41.81
$HF + HNO_3$	None	Original			Original	
NaOH (1%)	90.08	9.28	3.28	59.56	18.86	19.67
NaOH (10%)	60.24	14.21	24.69	8.72	32.37	57.30
KOH (1%)	93.52	1.90	1.20	83.86	6.95	8.06
KOH (10%)	75.04	7.88	14.96	13.95	30.37	52.76

Note. Original content of sample Nb — 39.65, Zn — 58.38,
$\Sigma(Nb + Zn)$ — 98.03%.

Fig. 3. Microsection containing the intermetallic phase
in the V—Zn system. Magnification 64.

A vanadium—zinc compound was observed on studying the corresponding microsections. The compound was never successfully separated, as it dissolved in acid, with the preferential dissolution of the zinc, as established by x-ray analysis. We see from Fig. 3 that the vanadium interacts completely with the zinc, forming an intermetallic compound. The change in microhardness from 120 to 298 kg/mm^2 indicates a wide range of homogeneity. The shapes of the precipitated particles are the same in all the microsections. These include individual, well-faced grains, mostly of quadrangular form (rhombs and squares), with regular triangles and hexagons, often having unequal sides. There are also crystals of skeletal growth form, and concretions comprising growth twins lying at right angles. The shapes of the grains suggest a cubic form of symmetry.

An x-ray study of the powder showed that vanadium zincide had the structure of copper with a lattice constant of 3.84 Å. The interatomic distance (2.72 Å) almost coincided with the sum of the atomic radii (2.73 Å). Rossteutscher and Shubert [6] found a compound VZn_3 with a cubic structure of the Cu_3Au type.

Comparing the structures of $NbZn_3$ and VZn_3 with our own niobium and vanadium zincites, we see that the resultant vanadium intermetallic compound lies in the same chemical and structural relationship to VZn_3 as $NbZn_{2.0-2.7}$ to $NbZn_3$, and that its structure constitutes the result of disordering in the VZn_3 compound attributable to a reduction in zinc content.

In view of the fact that the tantalum intermetallic compound proved inseparable, the formation of a compound between Ta and Zn had to be established by microhardness measurements, supported by the fact that a phase of high reflecting power appeared in the zinc. The Vickers microhardness of the compound was approximately 320-360 kg/cm^2 and that of tantalum 165 kg/cm^2.

Conclusions

1. We have obtained a new intermetallic compound of niobium and zinc with a variable composition $NbZn_{2.0-2.7}$ and have determined its structure and lattice constant.

2. We have found a compound in the V–Zn system with the structure of copper, having a wide range of homogeneity, and have determined its lattice constant and approximate composition.

3. We have observed an ordering–disordering phenomenon in the Nb–Zn and V–Zn systems.

4. We have proposed a method of chemical analysis for niobium zincide and studied its solubility in alkalis and acids.

Literature Cited

1. W. F. Hillebrand, N. E. Lendel, and G. A. Breit, Practical Handbook on Inorganic Analysis [Russian translation], Goskhimizdat, Moscow (1960), p. 441.
2. V. N. Gurin et al., Soviet Patent No. 172,285, dated July 3, 1964.
3. Yu. N. Knipovich and Yu. V. Morachevskii (editors), Analysis of Raw Minerals [in Russian], Goskhimizdat, Moscow (1959), pp. 472, 660, 663.
4. A. Martin, B. Knigton, and H. M. Feder, J. Chem. Eng., 6(4):596-599 (1961).
5. Protecting Columbium with Zinc, a Progress Report, Metal Progress, 81(3):172, 174-175 (1962).
6. W. Rossteutscher and K. Shubert, Z. Metallkunde, 55(10):617 (1964).
7. W. Rossteutscher and K. Shubert, Z. Metallkunde, 56(10):730 (1965).
8. J. Sandoz, J. of Metals, 12(4):340 (1960).
9. K. Shubert, Kristallstrukturen zweikomponentiger Phasen (1964).
10. C. L. Vold, Acta Cryst., 13:743 (1960).
11. C. L. Vold, Acta Cryst., 14:1289-1290 (1961).

DETERMINATION OF TUNGSTEN IN
BINARY TUNGSTEN—MOLYBDENUM ALLOYS

Z. S. Mukhina, L. I. Il'ina, and N. S. Kondukova

Moscow

The determination of tungsten in molybdenum is a complex analytical problem. In order to separate the metals the molybdenum is precipitated by hydrogen sulfide in the presence of tartaric acid. The tungsten is determined in the filtrate [2]. If there is only a little tungsten present the error is considerable.

For the gravimetric determination of tungsten, methods based on organic precipitating agents are now recommended. Thus Chernikhov and Goryushina [9] and Gusev and Kumov [4], use pyramidon for precipitating tungsten, Golubtsova [3] uses β-naphthoquinoline for the gravimetric determination of tungsten in steels and in ferro-tungsten. Other agents recommended include rivanol, methylene blue, and nicotine [5]. A method of determining tungsten with hydroxyquinoline in the presence of Complexone III has been described [7]. In the ASTM standards for 1965, cinchonine with a trace of benzoinoxime is recommended for the gravimetric determination of tungsten.

For the colorimetric determination of tungsten and molybdenum the thiocyanate method is employed [2, 6], this being based on the formation of thiocyanate complexes of molybdenum and tungsten at different acidities of the solution, and also the dithiole method [2, 8, 10].

Hexavalent molybdenum is completely extracted by dithiole with petroleum ether from a medium with an acidity of 6-14 N with respect to H_2SO_4. Tungsten forms an analogous complex with dithiole, only in a medium 0.5-2 N with respect to H_2SO_4 [11, 12].

Molybdenum may be separated from tungsten by extraction with acetyl acetone in a 6 M solution of sulfuric acid and also with a mixture of chloroform and acetyl acetone in 2 N hydrochloric acid [13].

Recently sulfur-containing organic reagents forming low-solubility compounds with metals of the hydrogen sulfide group, namely, thioacetamide, thionalide, thioxine, bismuthiol, diethyl-dithiocarbaminate, dithizone, and others have come into analytical practice in place of hydrogen sulfide (a poisonous gas). These are used for the precipitation of elements or extraction [1, 8, 10].

We considered two modes of separating molybdenum and tungsten: the precipitation of molybdenum with thionalide and the extraction of molybdenum with a mixture of chloroform and acetyl acetone.

TABLE 1. Determination of Tungsten in
Tungsten–Molybdenum Alloys

Alloy mixture		Tungsten deter- mined, %	Relative error, %	Method of separating the Mo and W	Method of determination
tungsten	molyb- denum				
75.0	25.0	74.4	−0.8	Precipitation with thionalide	Gravimetric with β-naphthoquinoline
75.0	25.0	75.1	+0.1	The same	The same
50.0	50.0	49.8	−0.4	» »	» »
50.0	50.0	48.7	−2.6	» »	» »
50.0	50.0	49.3	−1.4	» »	» »
50.0	50.0	50.7	+1.4	» »	Polarographic
25.0	75.0	24.0	−3.2	» »	Gravimetric with β-naphthoquinoline
25.0	75.0	24.3	−2.8	» »	Polarographic
25.0	75.0	24.6	−1.6	» »	Photocolorimetric
25.0	75.0	24.3	−2.8	Extraction	The same
3.0	97.0	2.85	−5.0	»	» »
3.0	97.0	2.90	−3.3	»	» »
1.5	—	1.5	0.0	»	» »
1.5	—	1.44	−4.0	»	» »
1.5	—	1.58	+5.3	»	Polarographic

Thionalide (structure) —NH—CO—CH$_2$—SH precipitates many elements in solutions of mineral acids: arsenic, antimony, tin, gold, palladium, copper, silver, mercury, and bismuth. In an ammoniacal solution in the presence of the complexing agents only thallium is precipitated. There are no data relating to the precipitation of tungsten and molybdenum in the literature. According to our own observations, molybdenum is completely precipitated by thionalide in a mineral-acid medium at pH 1–3. The complete precipitation of molybdenum is achieved in the presence of a twofold excess of thionalide.

Tungsten fails to enter into any reaction with thionalide and remains in solution; it may be determined in the solution by gravimetric or polarographic methods (Table 1). This method of separating molybdenum and tungsten gives the most accurate results for alloys containing 25% or more of tungsten (Table 2).

A sample 0.1–0.2 g in weight is placed in a platinum crucible and treated with a mixture of hydrofluoric and nitric acids. The solution in the crucible is evaporated to dryness, the dry residue being treated with nitric acid twice and water once, each time repeating the operation of evaporating to dryness. The dry residue is fused with 1.5 g of sodium carbonate, the melt is leached in water and diluted with water to a volume of 150–200 ml, a pH of 1 is established with hydrochloric acid, and 1–3 ml of a 10% solution of ammonium persulfate are added with 30–40 ml of a 2% acetic acid solution of thionalide. The residue is shaken vigorously, the solution with the residue is transferred to a 250-ml measuring flask, and this is brought up to the mark with water, agitated again, and left for 30 min in order to settle the precipitate. The solution and precipitate are filtered through a dry filter into a dry beaker. In the gravimetric determination of tungsten, an aliquot portion of the solution is taken (50 and 100 ml) and heated to 60–70°C, when the tungsten is precipitated with 30 ml of a 1% solution of β-naphthoquinoline. The solution with the deposit is held in a water bath to coagulate the precipitate,

TABLE 2. Determination of Tungsten by the Amperometric Method

Taken, g	Found, g
0.034	0.0336
0.052	0.053
0.026	0.0258
0.068	0.069
0.039	0.040
0.014	0.0144

the latter is filtered off, washed two or three times with a 1% solution of HCl, placed in a weighed porcelain crucible, dried, and roasted at 500-600°C.

In the polarographic determination of tungsten, an aliquot portion of solution (10 ml) is placed in a heat-resistant beaker and the organic material is decomposed by a mixture of sulfuric and nitric acids, the result then being evaporated until a light-yellow deposit is obtained. The deposit is dissolved in strong HCl and the tungsten is determined polarographically with a base electrolyte comprising an 8 N solution of HCl. The potential of the tungsten half wave is − 0.42 V.

The molybdenum may be extracted with a mixture of chloroform and acetyl acetone from a less acid medium than that required for extraction by the expensive reagent acetyl acetone on its own.

An alloy sample 0.5 g in weight is placed in a platinum crucible and treated with a mixture of HF and HNO_3. The solution in the crucible is evaporated to dryness, the dry residue being twice treated with nitric acid and once with water, each time repeating the evaporation to dryness. The dry residue is fused with 3 g of sodium carbonate. The melt is leached with water, transferred to a 100-ml measuring flask, and brought up to the mark with water; 20 ml of the resultant solution are transferred to a separating funnel, the appropriate amount of HCl is added to achieve a 2 N state, 20 ml of the organic mixture (acetyl acetone + chloroform, 1:1) are added, and extraction proceeds for 3 min. After the phase separation the lower organic layer is run off into another separating funnel and the aqueous layer is transferred to a beaker. The organic phase is washed with 30 ml of 2 N HCl after the extraction of the molybdenum, the organic layer is decanted, and the aqueous layer is transferred to the beaker containing the main aqueous phase and boiled for 5-10 min in order to remove organic material from the solution.

If the alloy contains more than 5% of tungsten, the determination is completed gravimetrically. For this purpose the aqueous layer is diluted with water to 100-150 ml and the tungsten is precipitated by heating with a 1% solution of β-naphthoquinoline as described earlier.

For a tungsten content of under 5% the aqueous layer is boiled for 5 min after the extraction of the molybdenum in order to remove organic substances, transferred to a 100-ml measuring flask, an aliquot portion is taken, and the tungsten is determined photocolorimetrically with an FÉK-N-57 photocolorimeter in the form of a thiocyanate complex. The optical density of the thiocyanate complex of tungsten is measured in cuvettes with a length of $l = 30$ mm, using a No. 2 light filter (blue). The reducing agent employed is a freshly prepared solution of titanium trichloride.

In the polarographic determination of tungsten, after the extraction of the molybdenum the aqueous layer is placed in a 50- to 70-ml beaker, evaporated to dryness, and the residue is treated with a mixture of sulfuric and nitric acids in order to remove organic material. The residue is dissolved in 25 ml of strong hydrochloric acid, transferred to a 100-ml measuring flask, and brought up to the mark with water. Depending on the tungsten content of the sample, an aliquot portion of the solution is taken so as to contain not more than 10 mg of tungsten, and the latter is determined polarographically with a base electrolyte comprising an 8 N solution of hydrochloric acid. The tungsten half-wave potential is − 0.42 V.

β-Naphthoquinoline may be used not only for the gravimetric but also for the amperometric determination of tungsten. Even in 1946 W. Sandberg developed a method for the amperometric titration of cadmium with β-naphthoquinoline in a 0.5 M solution of sulfuric acid in the presence of potassium iodide at a potential of − 0.9 V.

We tried using β-naphthoquinoline as a titrating solution for determining large quantities of tungsten. Titration took place at pH 3.5 in solutions of sodium tungstate. In order to acidify

TABLE 3.
Determination of
Tungsten in Alloys

Content in %	Found %	
	Thiocya-nate method	Ampero-metric method
4.50	4.54	4.48
5.60	5.74	5.66
10.20	10.05	10.12
15.40	15.12	15.29
20.60	20.76	20.46

these to the necessary pH value, solutions of phosphoric acid were employed, these forming easily-soluble complex phosphorotungstates with tungsten. Migration currents were suppressed by potassium iodide. Amperometric titration was applied to solutions of sodium tungstate with a known tungsten content (Table 3). In all cases the volume of the solution for titration was equal to 40 ml. To each sample 2 g of potassium iodide were added and the result was titrated in the cold with a solution of β-naphthoquinoline at a voltage of -0.86 V.

Apart from artificial mixtures, alloys containing 0.5-60% of tungsten together with titanium, aluminum, iron manganese, and chromium were analyzed.

An alloy sample 0.5 g in weight is dissolved in 40 ml of sulfuric acid (1:3) while heating. On dissolution of the sample, nitric acid is added drop by drop until the liquid turns yellow and the result is evaporated until SO_3 appears. The residue is cooled and dissolved in 50 ml of water with a little heating. The acid solution is transferred to a 500 ml measuring flask containing 30 ml of 20% caustic soda and agitated. After cooling, water is added up to the mark and the result is again shaken. The solution is filtered through a dry filter into a dry beaker, and for determination purposes 10-15 ml of the solution are taken, neutralized with dilute (1:10) phosphoric acid, diluted with water to a volume of 35-40 ml, 2 g of potassium iodide are added, and the result is titrated with a 0.5% solution of β-naphthoquinoline.

Conclusions

1. We have developed methods of determining tungsten in alloys of the tungsten-molybdenum type. For a tungsten content of over 25% we use a method based on the precipitation of molybdenum with thionalide and the subsequent gravimetric and polarographic determination of tungsten; for 1.5% of tungsten or over we use a method based on the extraction of molybdenum with a mixture of acetyl acetone and chloroform (1:1) and subsequent gravimetric and polarographic determinations.

2. The accuracy of the gravimetric and polarographic methods (tungsten content 25% or over) is ±1.5%, that of the polarographic and photocolorimetric methods (tungsten content under 25%) is ±3% (relative in both cases).

3. We have demonstrated the possibility of using an amperometric method of determining tungsten by titration with a solution of β-naphthoquinoline; the accuracy of the method is ±2% (relative) for a tungsten content of 0.5-20%.

Literature Cited

1. A. I. Busev, Analytical Chemistry of the Elements — Molybdenum [in Russian], Izd. AN SSSR, Moscow (1963).
2. W. F. Hillebrand et al., Practical Handbook on Inorganic Analysis [Russian translation], Goskhimizdat, Moscow (1957).
3. R. B. Golubtsova, Zh. Analit. Khim., 3:148 (1948).
4. S. I. Gusev and I. I. Kumov, Zh. Analit. Khim., 3:374 (1948).
5. M. N. Isaeva, Transactions of the Institute of Oceanology [in Russian], Vol. 47, AN SSSR (1961), pp. 157-181.
6. Z. S. Mukhina et al., Methods of Analyzing Metals [in Russian], Oborongiz, Moscow (1959).
7. R. Pribil, Complexones in Chemical Analysis [Russian translation], IL, Moscow (1960).
8. E. V. Sendel, Determination of Traces of the Elements [Russian translation], IL, Moscow (1964).

9.   Yu. A. Chernikhov and V. G. Goryushina, Zavod. Lab., 11:136 (1945).
10. H. Charlot, Methods of Analytical Chemistry. Quantitative Analysis of Inorganic Com-
     pounds [Russian translation], Khimiya, Moscow (1965).
11. S. H. Allen and M. B. Hamilton, Analyt. Chim. Acta, 7:483 (1952).
12. J. P. Koveney and H. Treiser, Analyt. Chim. Acta, 29:290 (1957).
13. R. Weiner and B. Bariss, Z. analyt. Chem., 168:195 (1959).

DETERMINATION OF MOLYBDENUM IN
THE PRESENCE OF TUNGSTEN

V. G. Shcherbakov, Z. K. Stegendo, and R. A. Antonova

All-Union Scientific-Research Institute of Hard Alloys

The separate determination of large quantities of molybdenum and tungsten in their mutual presence is a complicated analytical problem. Existing methods take a long time and require subsequent colorimetric determination of each component. By the acid hydrolysis of tungsten, complete separation of these elements is never achieved even from a sulfuric acid medium.

We found that the separation of tungsten and molybdenum with β-naphthoquinoline was also restricted to a limited molybdenum content for a high acidity of the solution; above this content the molybdenum was coprecipitated with the tungsten.

Regarding the determination of molybdenum in the presence of tungsten, some work by Gopala Rao* is of particular interest; this author first found the redox potential of the pair Mo^{6+}/Mo^{5+} in phosphoric acid of various concentrations.

A comparison of the potentials of the Mo^{6+}/Mo^{5+} and Fe^{3+}/Fe^{2+} pairs showed that in orthophosphoric acid up to a concentration of 5.8 M the Fe^{3+}/Fe^{2+} pair had a much higher redox potential than Mo^{6+}/Mo^{5+}. For an orthophosphoric acid concentration of over 11 M, the difference between the potentials of Mo^{6+}/Mo^{5+} and Fe^{3+}/Fe^{2+} and the orthophosphoric acid concentration is indicated in Fig. 1.

For an orthophosphoric acid concentration of about 13 moles the difference in the potentials between the two pairs is considerable; hence in 13 M solutions of orthophosphoric acid hexavalent molybdenum rapidly oxidizes divalent iron. The redox potential of Mo^{6+}/Mo^{5+} in a 13 M solution of orthophosphoric acid equals 0.685 V and that of Fe^{3+}/Fe^{2+} 0.386 V; the transition potentials of methylene blue and thionine are 0.55 and 0.56 V respectively. These indicators may thus be used for titrating molybdenum with a solution of Mohr's salt.

The molybdenum in solutions of known Mo content was determined in the following way. A solution containing 1–20 mg of molybdenum was diluted with water to a volume of 30 ml and treated with 60 ml of orthophosphoric acid and 1 ml of a 0.1% solution of methylene blue before being titrated with a 0.05 M solution of Mohr's salt in a current of CO_2. After each addition of two drops of the titrated solution of Mohr's salt, the solution was carefully agitated for 15 sec and then more titrated Mohr's salt solution was added. In titrating small amounts of molybde-

*G. Gopala Rao, Talanta, 10:1047 (1963).

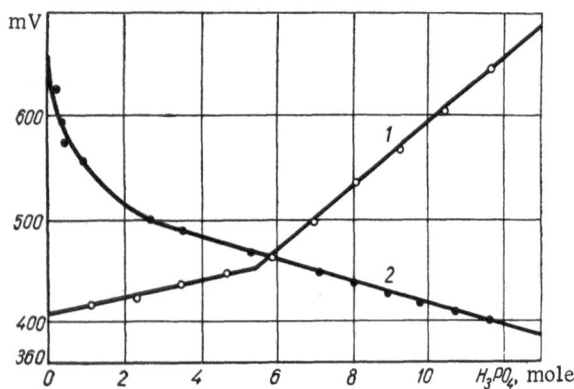

Fig. 1. Electrokinetic potential of Mo^{6+}/Mo^{5+} (1) and Fe^{3+}/Fe^{2+} (2) as a function of orthophosphoric acid concentration.

num (1-10 mg) the solution became pale blue close to the end of the titration and lost its color at the end point. For molybdenum contents of 10-20 mg the coloring of the solution, blue at the beginning of the titration, became pale green at the end. The results of the experiments are shown in Table 1. The error in the determination was no greater than 9% (relative) in determining small quantities of molybdenum (around 3 mg); for Mo contents of over 3 mg the error rarely exceeded 3%.

In determining molybdenum in solutions with different Mo:W ratios (Table 2) the test solution was diluted with water to 30 ml and then the process continued as in the determination of molybdenum without tungsten. For a molybdenum content of up to 5 mg the relative error was no greater than 6%; for a content of up to 24 mg it was 4%.

In order to determine the effect of sulfuric and hydrochloric acids on the determination of molybdenum, 5 ml of these acids were added to the solutions containing tungsten and molybdenum in addition to the orthophosphoric acid. The results presented in Table 3 showed that sulfuric and hydrochloric acids had no harmful effect on the titration of molybdenum.

A reduction in the phosphoric acid concentration makes the titration less sharp, ammonium salts have no effect, but the presence of chromium, manganese, and vanadium in large quantities does interfere with the determination of molybdenum.

TABLE 1. Determination of Molybdenum by Titration with a Solution of Ferrous Sulfate

Mo taken, mg	Mo found, mg	Difference, mg	Relative error %
1.02	0.97	—0.05	—4.9
1.58	1.44	—0.14	—8.9
2.52	2.74	+0.22	+8.7
3.16	3.28	+0.12	+3.8
4.74	4.80	+0.06	+1.2
6.30	6.40	+0.10	+1.5
7.90	8.00	+0.10	+1.2
11.06	11.36	+0.30	+2.7
11.98	11.84	—0.14	—1.1
15.80	15.55	—0.25	—1.5
19.00	19.11	+0.11	+0.6
19.50	19.56	+0.06	+0.3
20.00	19.84	—0.16	—0.8

TABLE 2. Volumetric Determination of Molybdenum in the Presence of Tungsten (69.6 mg)

Mo taken, mg	Mo found, mg	Error of the determination mg	%
1.02	0.97	0.05	—5.0
2.27	2.32	0.05	+2.2
3.00	3.20	0.20	+6.6
3.47	3.68	0.21	+6.0
5.53	5.36	0.17	—3.0
7.90	8.00	0.10	+1.2
7.94	7.99	0.04	+0.5
11.85	12.30	0.45	+3.8
14.22	14.76	0.54	+3.8
15.80	15.40	0.40	—2.5
15.80	15.90	0.10	+0.6
17.38	17.83	0.45	+2.5
19.75	20.20	0.45	+2.2
22.23	22.14	0.09	—0.4
23.82	23.98	0.16	+0.6

TABLE 3. Determination of
Molybdenum in the Presence of
Sulfuric or Hydrochloric Acids
(Tungsten Content 69.6 mg
in All Experiments)

Mo taken, mg	Acid	Mo found, mg	Differ-ence, mg	Relative error, %
3.00	HCl	3.06	0.06	+2.0
5.53	»	5.60	0.07	+1.2
7.90	»	8.12	0.22	+2.7
11.91	»	12.00	0.08	+0.7
13.43	»	13.58	0.15	+1.1
15.80	»	15.64	0.16	−1.0
3.95	H_2SO_4	4.10	0.05	+1.3
7.90	»	8.37	0.47	+5.8
11.91	»	12.00	0.08	+0.7
15.80	»	15.80	0	0

TABLE 4

Mo taken, mg	Mo added, mg	Mo found, mg	Error of the determination	
			mg	rel.%
9.65	28.42	9.60	0.05	−1.0
9.65	48.63	9.60	0.05	−1.0
19.30	28.42	19.20	0.1	−0.5
28.95	42.63	28.80	0.15	−0.5
48.25	42.63	48.48	0.23	+0.5

TABLE 5. Determination
of Molybdenum in
Tungsten Concentrates, %

Tungsten content	Method		
	gravi-metric	volu-metric	potentio-metric
50.2	4.90	4.80	4.85
16.5	2.66	2.74	2.60

Molybdenum may also be determined by potentiometric titration in a 13 M solution of phosphoric acid. The indicator electrode is a platinum wire and the comparison electrode a standard calomel. The agitation of the solution accelerates the attainment of the equilibrium potential. The conditions of titration are the same as in the volumetric determination.

A sharp jump of potential takes place on titrating molybdenum up to 20 mg; if more than this is present no sharp potential jump occurs, and the end of the titration can only be judged from a slow change near the balance point. In order to locate the potential jump at the balance point in this case 1 ml of methylene blue indicator is introduced into the solution.

The indicator loses its color a little before the equilibrium potential is reached. The presence of sulfuric or hydrochloric acid in the solution·has no effect on the titration. We see from the data presented in Table 4 that the potentiometric titration of molybdenum in the presence of tungsten gives excellent results.

Table 5 indicates the determination of molybdenum in two tungsten concentrates as obtained by the gravimetric, volumetric, and potentiometric methods.

Conclusions

1. We have developed volumetric and potentiometric methods of determining molybdenum in the presence of tungsten, based on the redox reaction of Mo^{6+}/Mo^{5+} and Fe^{3+}/Fe^{2+} in a 13 M solution of orthophosphoric acid.

2. The method thus developed enables us to determine molybdenum in the presence of tungsten in various solutions and also in ores and concentrates.

PHOTOMETRIC DETERMINATION OF BORON IN NICKEL AND TITANIUM BORIDES USING MAGNEZONE I IN AN ALKALINE MEDIUM

E. I. Nikitina

Moscow

In order to determine boron in the borides of refractory metals as well as in ores, minerals, and steels containing more than 0.2% of this element, a classical volumetric method of titration with a solution of caustic soda in the presence of glycerin or mannitol is often employed [1, 2, 6]. For determining boron in quantities of under 0.05% colorimetric methods are used; these incorporate quinalizarin, carminic acid, arsenazo I, and so on, the reaction with boron taking place in concentrated sulfuric acid [1, 4, 7].

Some new organic reagents for determining boron from weakly-acid solutions have also been studied [3, 5]. N. S. Poluéktov [7] indicates that boron should have color reactions in weakly-acid and alkaline media with organic reagents of the hydroxyazo dyes having OH groups in the para position with respect to the azo group in one ring and in the ortho position in the other.

For the colorimetric determination of boron in an acetic acid medium, N. S. Poluéktov proposed using n-resorcinol. Hardly any method has been proposed for the colorimetric determination of boron in alkaline media. In this paper we shall describe a method for the photocolorimetric determination of boron with Magnezone I $\left(O_2N-\left\langle\right\rangle-N=N-\left\langle\right\rangle-OH\right)$ in a weakly-alkaline medium.

At pH 8-10.5 such solutions have a brownish-red color, becoming violet at pH 12-13. We found that boron formed a yellow compound with Magnezone I in a weakly-alkaline medium. The effect of the pH of the solution on the coloring of the compound of boron and Magnezone I is shown in Fig. 1, from which we see that the reaction between boron and Magnezone I with the formation of a colored complex only occurs in a weakly-alkaline medium at pH 8-10.5. The

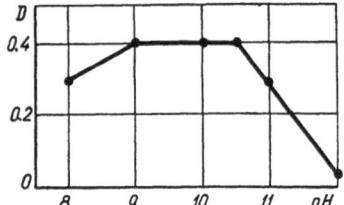

Fig. 1. Effect of pH on the optical density of solutions of a compound of boron with Magnezone I.

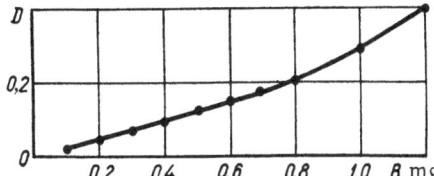

Fig. 2. Calibration curve for determining boron with Magnezone I (0.1-1.2 mg of B in 100 ml of solution).

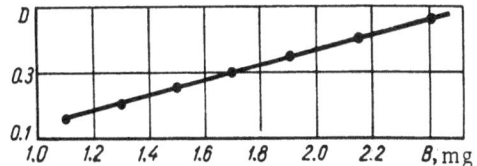

Fig. 3. Calibration curve for determining boron with Magnezone I (1.1-2.4 mg of boron in 100 ml of solution).

colored compound of boron with Magnezone I is formed at room temperature immediately after adding the reagent in question to a weakly-alkaline solution of boron. The coloring of the compounds of boron with Magnezone I is stable, easily reproducible, and entirely suitable for photometric determination. For a boron content of 0.11-3.0 mg the coloring of the compound so formed obeys the Lambert–Beer law (Figs. 2 and 3). Solutions of Magnezone I in an alkaline medium are strongly colored, and different amounts of solution are accordingly used for determining different boron concentrations. For the photometric determination of 0.2-1.2 mg of boron in 10 ml of solution a calibration curve is plotted with 6 ml of a 0.05% solution of Magnezone I (Fig. 2), for 1.0-2.8 mg with 12 ml (Fig. 3), and for 2.8 mg with 16 ml of the reagent.

Figure 4 presents the light-absorption curves of solutions of Magnezone I and the compound formed between the latter and boron. The intensity of the coloring of the boron compounds was measured at 560 mμ in the region of the yellow light filter.

Table 1 presents some data relating to the photocolorimetric determination of boron with Magnezone I in a weakly-alkaline medium, indicating the excellent reproducibility of the results obtained by this method.

Since the reaction of boron with Magnezone I takes place in an alkaline medium, in its practical application to alloys it is essential to eliminate components forming hydrates of metal oxides in alkaline media, or to combine these into a complex.

TABLE 1. Effect of Other Metals on the Reaction of Boron with Magnezone I

Introduced, mg			B taken, mg	B found, mg
Al	Mo	Ba		
0.5	2.0	0.1	0.54	0.56
0.5	—	0.1	1.29	1.25
—	2.0	0.1	2.16	2.11
—	2.0	0.1	1.94	1.9
0.5	—	0.1	5.0	3.2
—	2.0	0.1	3.0	3.1

Effect of Complexing Substances on the Reaction of Boron with Magnezone I

The foregoing investigation showed that the presence of tartaric, oxalic, acetic, carbonic, and nitric acids in the solution as well as that of hydrogen peroxide and other oxidizing agents, interfered with the determination of boron by means of Magnezone I, while the ions of hydrochloric and sulfuric acids had no such effect.

We also studied the effect of various cations remaining in the dissolved state in alkaline media during the analysis of borides on the reaction between boron and Magnezone I; these included

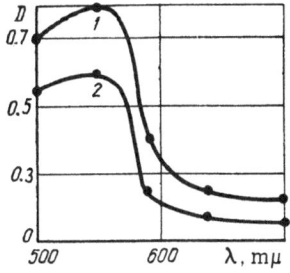

Fig. 4. Light-absorption curves of Magnezone I (1) and its compound with boron (2).

molybdenum, aluminum, and barium (Table 1). The presence of 0.5 mg of aluminum or barium, or up to 2 mg of molybdenum in 100 ml of solution had no harmful effects.

Analytical Procedure

To a solution of boric acid in a 100-ml measuring flask we add one drop of a 0.01 M methyl red indicator and neutralize it with a 1 N solution of sulfuric acid or a 0.1 N solution of caustic soda until the color turns light red. The flask is heated to boiling, then a 0.1 N solution of caustic soda is added carefully, drop by drop, from a burette until the rose color vanishes. To the neutralized solution we add 6-12 ml of a 0.03% alkaline solution of Magnezone I from a burette, and immediately afterwards 5 ml of a 0.5% solution of gelatin, agitate the mixture, bring up to the mark with freshly boiled water, and agitate again.

The intensity of the coloring is measured in a photocolorimeter with a yellow light filter, using a cuvette length of 50 mm for 6 ml of reagent and 20 mm for 12 ml. The boron content is found from a calibration chart:

Boron taken, mg	Boron found, mg	Boron taken, mg	Boron found, mg
0.11	0.11	1.94	1.95
0.32	0.32	2.16	2.16
0.54	0.53	2.4	2.3
0.64	0.64	3.0	3.1
1.08	1.1	3.5	3.45
1.29	1.3		

Determination of Boron in Nickel Borides

The color reaction of boron with Magnezone I taking place in a weakly-alkaline medium was used for the photocolorimetric determination of boron in nickel, titanium, and chromium borides.

The nickel borides contain up to 30% boron; depending on the method of production they may or may not dissolve in hydrochloric acid. In order to determine the boron in borides soluble in hydrochloric acid for a boron content of 0.5-2.5% we take samples of 1 g, for large contents 0.2 g.

The sample is dissolved in 50 ml of hydrochloric acid (1:1) while heating (not boiling) for 2-3 h. After dissolution, five drops of a 3% solution of hydrogen peroxide are added and boiled for 15 min. The solution is cooled, transferred to a 200-ml measuring flask, and the excess of acid is neutralized with a 20% solution of caustic soda until a deposit forms. Then a 2-ml excess of caustic soda is added (the solution over the deposit should be colorless), water is added up to the mark, and the mixture is shaken. The nickel then passes quantitatively into the residue. Some of the solution is filtered, 20 ml of solution is transferred to a 100-ml measuring flask, one drop of methyl red indicator is added, and this is followed by the drop-by-drop addition of a 1 N solution of sulfuric acid until a bright red color is achieved. The flask containing the solution is heated to boiling and boiled for 5 min. Then a 0.1 N solution of caustic soda is added to the solution from a burette until the red color vanishes. The accuracy of the neutralization is verified by adding one drop of a 1 N solution of sulfuric acid until a red color appears and then again carefully adding a 0.1 N solution of caustic soda drop by drop until the solution becomes yellow. To the neutralized solution we add 6-12 ml of a solution of Magnezone I and then immediately 5 ml of a 0.5% solution of gelatin, shake the mixture, add water (freshly boiled) up to the mark, and shake again. The intensity of the coloring is measured in a photocolorimeter as before. The boron content is determined from a calibration curve. A dummy experiment is carried out at the same time with all the reagents involved and the optical-density

TABLE 2. Comparative Data Relating to the Determination of Boron with Magnezone I and by the Volumetric Method in Nickel Borides, %

Experiment No.	Sample weight, g	Boron found volumetrically	Boron found photometrically	
1	1.0	1.08	1.07	1.11
2	1.0	1.2	1.29	1.25
3	1.0	1.75	1.83	1.87
4	1.0	2.05	2.26	2.1
5	0.2	6.0	5.8	6.08
6	0.2	8.0	8.1	8.2
7	2.0	0.38	0.42	0.44
8	0.1	9.0	9.2	9.4
9	0.1	12.5	12.1	12.3
10	0.1	22.4	22.3	22.5

TABLE 3. Comparative Data Relating to the Determination of Boron in Titanium Borides, %

Experiment No.	Sample weight, g	Method		
		volumetric	colorimetric (with Magnezone I)	
1	0.1 Ti + B	10.0	10.2	10.5
2	The same	20.0	20.5	—
3	0.1 g boride	18.9	19.1	18.8
4	The same	18.9	19.4	19.3
5	» »	24.9	25.1	25.4
6	» »	24.9	25.3	25.8
7	» »	28.9	28.5	28.7
8	» »	28.9	28.6	28.8
9	» »	30.0	29.5	29.8

reading of the dummy experiment is subtracted as a correction. The results of the photocolorimetric determination of boron with Magnezone I in nickel borides (Table 2) show that this method is valid for a boron content of 1-6% (experiments 1-7) and over (8-10). In the latter case the samples are baked in order to bring the alloy into the dissolved state.

The results of Table 2 indicate the excellent reproducibility of the photometric method and its agreement with the classical volumetric procedure involving titration with alkali. The method enables us to determine boron photometrically by means of the color reaction with Magnezone I for a content of 1-23%.

The reaction of boron with Magnezone I was also employed for determining boron in titanium borides and Ti-base alloys with chromium and molybdenum for boron contents of up to 25-29%. The titanium borides and alloys fail to dissolve in HCl, only dissolving in sulfuric acid in the presence of hydrogen peroxide; in addition to this they may be brought into the dissolved state by melting with sodium peroxide and caustic soda. In both cases the presence of oxidizing agent interferes with the reaction between boron and Magnezone I.

Determination of Boron in Titanium Borides Soluble in Acids

In order to determine the boron in titanium borides dissolved in sulfuric acid with hydrogen peroxide, we first found the conditions required to eliminate the effects of the oxidizing agents.

A titanium sample is dissolved in 20 ml of sulfuric acid (1:1), the level of the liquid in the beaker is noted, and in order to secure the complete dissolution of the material 6-10 ml of 30% hydrogen peroxide are added. The same quantities of reagent are poured into a beaker at the same time by way of a dummy experiment. The solutions are heated until the excess of hydrogen peroxide is removed (the titanium solution then loses its color) and evaporated to the mark previously noted so as not to produce any evolution of sulfuric acid vapor, the appearance of which denotes a loss of boron.

After the decomposition of all the sample, the solution is transferred to a 200-ml measuring flask, 10-15 ml of a 0.05 N solution of Mohr's salt are added (to reduce the traces of oxidizing agent remaining), and iron and titanium are precipitated with a 20% solution of caustic soda (not more than 2 ml). The solution in the measuring flask is brought up to the mark with water, shaken, and allowed to stand, a dummy experiment being conducted in parallel. For the colorimetric determination with Magnezone I 20 ml of the original solution are transferred to a 100-ml measuring flask and the analysis proceeds as described earlier. The amount of boron in the titanium boride is determined from a calibration curve. The corresponding results are shown in Table 3.

The results of Table 3 show that large quantities of boron (up to 28%) in titanium borides may be determined by the colorimetric method with Magnezone I. The error of the method when determining large quantities of boron is ±0.5-0.7%.

Existing titanium and nickel borides include some varieties not dissolving in acids, this being due to the method of producing the borides and their technological processing. Such materials must be melted, but without any oxidizing agents, since these interfere with the reaction between boron and Magnezone I. In such cases melting may well be replaced by baking the sample with barium carbonate. The titanium or nickel boride sample is placed in a tall porcelain crucible with 5 g of barium carbonate at the bottom. Some 10 g of barium carbonate are poured on the sample and carefully mixed with it; it is important to check that the walls of the porcelain crucible should be protected from the sample by a layer of barium carbonate. Then, after mixing, the sample is again covered on top with a layer of pure barium carbonate. Baking proceeds for 2-3 h at 1000°C in a muffle furnace. The temperature is accurately monitored with a thermocouple. After baking, the cake is dissolved in the cold by adding hydrochloric acid; then the HCl solution is neutralized with barium carbonate to give a weakly-alkaline reaction, transferred to a 200-ml measuring flask, brought up to the mark with water, and shaken. We then take 20 ml of the solution into a 100-ml measuring flask for the colorimetric determination of boron with Magnezone I as indicated earlier. A dummy experiment using all the reagents is carried out at the same time. The results of the experiments are shown in Table 3 (experiments 6, 8, 9) and 2 (experiments 8-10).

Conclusions

1. We have proposed a method for the photocolorimetric determination of boron with Magnezone I in an alkaline medium. The Magnezone I enables us to determine boron quantitatively for concentrations of 0.11 to 3.5 mg in 100 ml of solution in a weakly-alkaline medium at pH 9-10.5. We have found the optimum conditions for the photocolorimetric determination of boron, established the necessary pH of the solution, amount of reagent, reaction time, light absorption, temperature, and the effect of a number of cations on the reaction between boron and Magnezone I. We have found that the presence of up to 2 mg of molybdenum and 0.5 mg of aluminum and barium has no harmful effect. The sensitivity of the method is 1 μg of boron in 100 ml of solution.

2. We have developed the conditions for the photocolorimetric determination of boron with Magnezone I in nickel and titanium borides and their alloys.

The method enables us to determine from 1.0 to 28% of boron by the photocolorimetric process. The reaction of Magnezone I with boron enables us to determine large quantities of boron photometrically in an alkaline medium.

Literature Cited

1.   W. F. Hillebrand et al., Practical Handbook on Inorganic Analysis [Russian translation], Yu. Yu. Lur'e, ed., Goskhimizdat (1957).
2.   A. M. Dymov, Technical Analysis of Ores and Metals [in Russian], Metallurgizdat, Moscow (1949).
3.   G. V. Samsonov et al., Analysis of Refractory Compounds [in Russian], Metallurgizdat, Moscow (1962).

4.   Z. S. Mukhina and A. F. Aleshina, Zavod. Lab., No. 1 (1945).

5.   F. Faigel, Drop Analysis [Russian translation], Goskhimizdat, Moscow (1933).

6.   I. M. Korenman, Zh. Analit. Khim., 2(3):153 (1947).

7.   N. S. Poluéktov and N. P. Nikonov, in: Transactions of the Commission on Analytical Chemistry, Vol. 3, Izd. AN SSSR, Moscow (1951), p. 4.

SPECTROPHOTOMETRIC STUDY OF THE FORMATION OF COMPOUNDS BETWEEN NIOBIUM AND THE REAGENT PAN

V. I. Kornilova

Institute of Problems in Materials Science, Academy of Sciences of the Ukrainian SSR

The pyridine azo compounds first obtained by A. E. Chichibabin [1] have for a long time found no particular applications in analytical chemistry. In recent years these reagents have come to be used as complexonometric indicators and for the photometric determination of many elements. Of greatest interest are the azo compounds with hydroxy groups in the o position relative to the azo group, obtained by the combination of diazotized 2-aminopyridine with 2-naphthol, resorcinol, cresol, and other phenols. Most frequently employed are 1-(2-pyridylazo)-2-naphthol (PAN) and 1-(2-pyridylazo)-resorcinol (PAR):

It has been found [2] that, depending on the acidity, PAN may occur in solution in three forms: a yellow-green acid (H_2R^+) at pH $\leq$ 2, soluble in water with a solubility constant of $K_{dis} = 1.26 \cdot 10^{-2}$; a neutral yellow form (HR) at pH $\leq$ 2, insoluble in water and forming colloidal solutions; and a dissociated red form at pH 11, soluble in water with $K_{dis} = 6 \cdot 10^{-13}$. The dissociation of the reagent PAN may be expressed as follows [2]:

$$H_2R^+ \rightleftarrows HR + H^+; \quad HR \rightleftarrows R^- + H^+.$$

The most complete information regarding the use of pyridine azo compounds in analytical chemistry appears in a review by A. I. Busev and V. M. Ivanov [3]. Lassner and Plüachel [4] studied the reaction of peroxide complexes of niobium, tantalum, and titanium with certain metallochromic indicators, including PAN. It was found that at pH 1.5-7 niobium formed a red-violet compound extractable with amyl alcohol on reacting with PAN in the presence of hydrogen peroxide.

The aim of the present investigation is to make a more detailed study of the reaction underlying the formation of this compound.

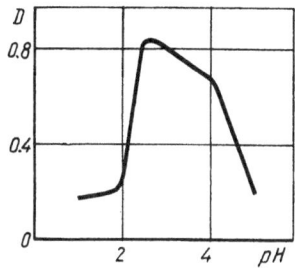

Fig. 1. Optical density of a solution of the compound of niobium with PAN as a function of the acidity of the solution (niobium concentration of $8 \cdot 10^{-5}$ mole/liter, PAN concentration of $1.6 \cdot 10^{-4}$ mole/liter, $\lambda = 560$ mμ, cuvette length 1 cm).

Preliminary experiments established that, on replacing hydrogen peroxide by other complexing agents (trilon B, tartaric, citric, and oxalic acids), no such colored compounds were formed.

The original solution of niobium was prepared by melting niobium pentoxide with potash and dissolving the resultant potassium niobate in water. The reagent PAN was prepared by dissolving an exactly weighed sample in ethyl alcohol.

The relation between the formation of the colored compound and the pH of the solution (varying from 1 to 5) is illustrated in Fig. 1. We see from the figure that the maximum optical density of the solutions of the complex compound occurs at pH 2.6, and that the complex is stable between pH 2.6 and 4.

We obtained the absorption spectra of the colored compound of niobium with PAN and also that of the PAN itself at pH 2.6 (Fig. 2). We see from this figure that the greatest difference in the optical densities of the complex and the reagent occurs at $\lambda = 560$ mμ. The optical density was subsequently measured at this wavelength.

We studied the optical density of the compound of niobium with PAN as a function of hydrogen peroxide concentration. The experiments were carried out as follows. We prepared a mixture from a calculated amount of potassium niobate and hydrogen peroxide. To this we added a calculated amount of alcohol and reagent and used a buffer solution with pH 2.6 to bring this up to a specified volume. The ratio of the niobium to the PAN in the test solutions was Nb:PAN = 1:2 in all cases. The hydrogen peroxide concentration was varied over a wide range (Fig. 3). We see from the graph that the introduction of hydrogen peroxide up to a ratio of $Nb:H_2O_2 = 1:250$ intensifies the coloring of the solution, but further increasing the concentration has no effect; only for $Nb:H_2O_2 = 1:2500$ is there a severe weakening of the color. Evidently the introduction of hydrogen peroxide up to a certain ratio of $Nb:H_2O_2$ helps the niobium to assume a reactive form.

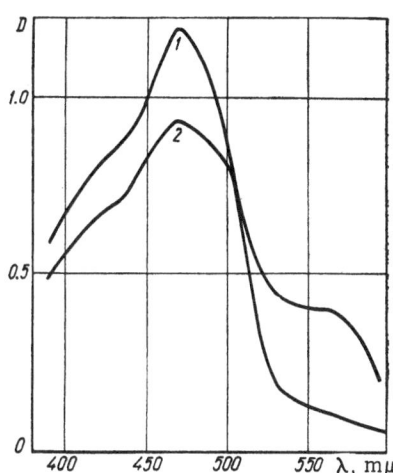

Fig. 2. Absorption spectra of solutions of PAN and its compound with niobium: 1) PAN at pH 2.6, concentration $8 \cdot 10^{-5}$ mole/liter; 2) compound of Nb with PAN at pH 2.6, niobium concentration $4 \cdot 10^{-5}$ mole/liter.

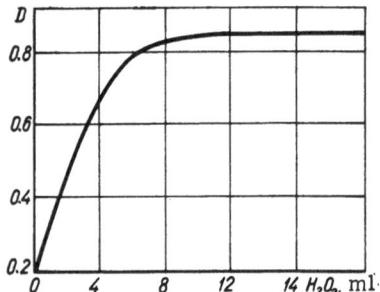

Fig. 3. Effect of an excess of hydrogen peroxide on the optical density of solutions of the Nb−PAN compound at pH 2.6 (niobium concentration is constant and equal to $8 \cdot 10^{-5}$ mole/liter, H_2O_2 concentration varies from $8 \cdot 10^{-5}$ to $2 \cdot 10^{-1}$ mole/liter).

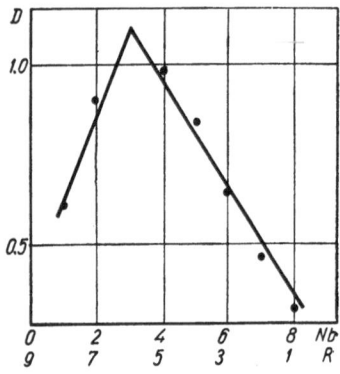

Fig. 4. Determination of the ratio between the reacting components by the method of isomolar series in the Nb−PAN system (total concentration of Nb and PAN $3.6 \cdot 10^{-4}$ mole/liter, $\lambda = 560$ mμ, cuvette 1 cm).

It is well known that a solution of PAN in an alkaline medium is colored red; the complex of niobium with PAN is also red; thus we may consider that in forming the colored complex the hydrogen atom of the hydroxyl group is simply replaced by the metal, which forms a more polar bond with the oxygen atom of this group than does the hydrogen.

The relation between the reacting components in the Nb−PAN system was determined by the method of isomolar series. The total concentration of the components was $3.6 \cdot 10^{-4}$ mole and the pH of the solution 2.6; all the solutions contained 0.1 ml of perhydrol. The optical density of the solutions was measured in an SF-4 spectrophotometer using a cuvette with a layer thickness of 1 cm. Data relating to the determination of the composition of the Nb−PAN compound are presented in Fig. 4; the maximum on the curve corresponds to the ratio Nb:PAN = 1:2.

The solutions of the colored Nb−PAN complex obey the Lambert−Beer law for niobium contents of 1.8-9 μg/ml.

Conclusions

We have found that niobium forms a red compound with an absorption maximum at $\lambda = 470$ mμ with PAN in acid solutions (pH 2.6-4) in the presence of hydrogen peroxide. The composition of the compound corresponds to a ratio of Nb:PAN = 1:2 between the reacting components.

Literature Cited

1.   A. E. Chichibabin and M. D. Ryazantsev, Zh. Ross. Fiz.-Khim. Obshch., 47:1582 (1915).
2.   B. F. Pease and M. B. Williams, Analyt. Chem., 31:1044 (1959).
3.   A. I. Busev and V. M. Ivanov, Zh. Analit. Khim., 19:1238 (1964).
4.   E. Lassner and R. Püachel. Mikrochim. Acta, 5:753 (1964).

DETERMINATION OF ARSENIC IN
HIGH-PURITY MOLYBDENUM

V. G. Shcherbakov and G. V. Onuchina

All-Union Scientific-Research Institute of Hard Alloys

The determination of small quantities of arsenic ($1 \cdot 10^{-4}\%$) in metallic molybdenum and its salts has never yet been described. The existing spectral method enables arsenic contents down to $5 \cdot 10^{-4}\%$ to be determined. The most sensitive and suitable method in the present case is the method of determining arsenic based on the formation of the heteropoly acid $H_3[As(Mo_3O_{10})_4]$ with the subsequent reduction of this to molybdenum blue. This method is used for determining arsenic in metallic tungsten [2].

An attempt to extend the method of determining arsenic in metallic tungsten to determining it in metallic molybdenum was unsuccessful. The aim of the present investigation was to work out the conditions for the photometric determination of arsenic in metallic molybdenum from the reaction involving the formation of an arsenic molybdic heteropoly acid.

In order to separate the arsenic from large quantities of molybdenum we attempted to coprecipitate it by means of ammonium with ferric hydroxide. This method gave no useful results. Nor did we have any success in attempts at separating phosphorus from arsenic by precipitation with $Ca(NO_3)_2$. Satisfactory results were obtained, however, on separating arsenic from phosphorus and molybdenum by the method proposed by Alekseev [1] based on the different solubility of the heteropoly acids of arsenic and phosphorus in organic solvents.

The phosphoromolybdic heteropoly acid was readily extracted by a mixture of isobutyl alcohol with chloroform. The arsenomolybdic heteropoly acid was extracted from an aqueous solution by a mixture of isopropyl alcohol, ethyl acetate, and chloroform.

We determined the optimum conditions for measuring arsenic contents of 0.001-0.005 mg experimentally. The optimum volume of the aqueous phase was 40 ml, the volume of the mixture of organic solvents 8 ml. Extraction took place for a solution acidity of 0.4 N referred to nitric acid. On maintaining these conditions, down to 0.5 μg of arsenic in 25 ml of solution could be determined. The optical density of the solution was measured in an FÉK-N photocolorimeter in a cuvette with a layer thickness of 50 mm, using a light filter No. 7 with an effective wavelength of 610 mμ.

The calibration curve was constructed as follows. To various quantities of a standard arsenic solution we added 10 ml of a 25% solution of sodium molybdate, neutralized this with nitric acid (sp. gr. 1.2) and introduced a 2.5-ml excess of the latter. In order to separate possible traces of phosphorus the solution was transferred to a separating funnel, diluted with water to 40 ml, shaken, and allowed to stand for 10 min in order to form the heteropoly acids; then

TABLE 1. Determination of Arsenic in a Standard Solution

Experiment No.	Arsenic taken, mg	Arsenic found, mg	Relative error, %
1	0.0005	0.00055	+10
2	0.0005	0.00035	−30
3	0.0010	0.00085	−15
4	0.0010	0.00095	−5
5	0.0020	0.0018	−10
6	0.0020	0.0019	−5
7	0.0030	0.0032	+7
8	0.004	0.0044	+10
9	0.005	0.0052	+4
10	0.006	0.0067	+12
11	0.007	0.0069	−1.5
12	0.009	0.0095	+6
13	0.010	0.0101	+1.0

TABLE 2. Determination of Arsenic in Samples of Metallic Molybdenum (Sample Weight 1 g)

Batch No.	Arsenic found		Mean arsenic content	
	mg	%	mg	%
261	0.00035	0.000035	0.0003	0.00003
	0.00025	0.000025		
259	0.0032	0.00032	0.003	0.0003
	0.0034	0.00034		
	0.0027	0.00027		
	0.0027	0.00027		
262	Not found			
257	0.0046	0.00046	0.0043	0.00043
	0.0039	0.00040		
255	Not found			
256	0.0030	0.00030	0.0029	0.00029
	0.0027	0.00027		

4 ml of isobutyl alcohol were added to saturate the aqueous solution, after which 4 ml of a mixture of isobutyl alcohol and chloroform (1:3 by volume) were added and the solutions were mixed by inverting the funnel 30 times. After allowing to stand, the lower layer was separated and the extraction was repeated once again. Under these conditions the phosphorus was extracted completely, this being verified by experiments with standard solutions of phosphorus.

After extraction of the phosphorus complex, 4 ml of a mixture of isobutyl alcohol and ethyl acetate (1:1 by volume) were added to the solution, which was shaken vigorously. Then chloroform (4 ml) was added and the solutions were mixed by inverting the funnel 30 times. After the separation of the layers, the lower layer was filtered into a 25-ml measuring flask. The extraction was repeated once again. The resultant yellow complex of arsenomolybdic heteropoly acid was reduced in the flask by 0.5 ml of a 0.4% solution of stannous chloride and diluted with isobutyl alcohol up to the mark. The optical density of the solution was measured on a photocolorimeter. We made a number of determinations of the arsenic in solution with a known

TABLE 3. Determination of Arsenic in Metallic Molybdenum in the Presence of an Added Standard Solution of Arsenic (Molybdenum Sample Weight 1 g)

Sample No.	Arsenic content of sample, g	Added arsenic, mg	Arsenic should be		Arsenic found to be		Relative error, %
			mg	%	mg	%	
262	Not found	0.001	0.001	0.0001	0.0011	0.00011	+10
	The same	0.002	0.002	0.0002	0.0017	0.00017	−15
	» »	0.003	0.003	0.0003	0.0033	0.00033	+10
	» »	0.004	0.004	0.0004	0.0042	0.00042	+5
	» »	0.005	0.005	0.0005	0.0048	0.00048	−4
	» »	0.006	0.006	0.0006	0.0061	0.00061	+2
	» »	0.008	0.008	0.0008	0.0080	0.00080	0
	» »	0.009	0.009	0.0009	0.0087	0.00087	−3
	» »	0.010	0.010	0.0010	0.0010	0.00110	+10
261	0.0003	0.002	0.0023	0.00023	0.0022	0.00022	−4
	0.0003	0.003	0.0033	0.00033	0.0030	0.00030	−9
258	0.003	0.001	0.004	0.0004	0.0042	0.00042	+5
	0.003	0.002	0.005	0.0005	0.0051	0.00051	+2
	0.003	0.004	0.007	0.0007	0.0075	0.00075	+7
	0.003	0.005	0.008	0.0008	0.0071	0.00071	−12
	0.003	0.006	0.009	0.0009	0.0093	0.00093	+3
	0.003	0.008	0.011	0.0011	0.0120	0.00120	+10

content. The method thus developed gave fairly accurate results (Table 1). Using the proposed method, we analyzed several batches of metallic molybdenum. The arsenic was determined under the following conditions. We dissolved 1 g of molybdenum over a water bath in 20 ml of aqua regia. The solution was evaporated to dryness, 10 ml of twice-distilled water were added, and evaporation to dryness was repeated. The molybdic acid was dissolved in 5 ml of a 20% solution of alkali, neutralized with nitric acid (sp. gr. 1.2), and a 2.5-ml excess of the latter was added. After this the solution was transferred to a separating funnel, diluted with water to 40 ml, and left for 10 min. Extraction proceeded in the same way as in the standard solution. In analyzing the samples, the molybdenum was used as a reagent for forming the arseno-molybdic complex (Table 2).

According to spectral analysis the arsenic content of these samples was under $5 \cdot 10^{-4}\%$. We see from Table 2 that the results of the arsenic determination are excellently reproducible. The accuracy of the method recommended for determining arsenic was verified by the method of additions. For this purpose a standard solution of arsenic was added to the sample in a specified ratio. The experimental results are shown in Table 3.

We see from the results presented in Tables 1-3 that the proposed method of determining arsenic gives a relative error of no more than 12%.

Conclusions

1. We have developed a photocolorimetric method of determining arsenic in metallic molybdenum. The method is based on the formation of arsenomolybdic blue and enables us to determine $1 \cdot 10^{-4}\%$ of arsenic.

2. The ultimate sensitivity is 0.5 μg of arsenic in 25 ml.

3. The relative error of the determination is no more than 12%.

4. The method here developed may be used to determine arsenic in molybdic anhydride, ammonium molybdate, and other molybdenum compounds.

Literature Cited

1. R. I. Alekseev, Zavod. Lab., No. 2 (1945).
2. V. G. Shcherbakov and Z. K. Stegendo, in: Hard Alloys [in Russian], Vol. 6, VNIITS, Moscow (1965).

CHEMICAL PHASE ANALYSIS OF MIXTURES OF BORIDES, CARBIDES, AND BOROCARBIDES

N. V. Vekshina and L. Ya. Markovskii

State Institute of Applied Chemistry

Chemical phase analysis has been widely used in experiments on metal–boron [1, 2] and metal–boron–carbon systems [3-7], and has led to the discovery of a large number of new and hitherto unknown individual phases. This method is based on a careful study of the chemical properties of the borides, in particular their chemical stability with respect to water and acids; it enables us to take the mixtures obtained by synthesis and extract from these various individual chemical compounds, the individuality of the latter subsequently being verified by x-ray phase analysis.

Our present aim is to demonstrate the effectiveness of this method by giving some specific examples.

Chemical Phase Analysis of Various Boride Systems

Earlier investigations [8-10] showed that borides have a lower hydrolytic stability and give a greater yield of boranes the greater their metal content. This enables us to distinguish various boride phases in the systems formed by the alkaline-earth and rare-earth metals with boron and to determine the proportions of these in test samples. In the synthesis of the borides in question, the powdered metals are mixed with boron and roasted at 800-1300°C. We obtained all the boride phases of the metals in question in this manner and studied their properties. The tetraborides of La, Ce, and Gd hardly dissolve at all in very dilute HCl, nor do the hexaborides of these metals or calcium. In concentrated HCl (at boiling point) only the tetraborides dissolve. The hexaborides dissolve readily in nitric acid. On further studying the hydrolytic decomposition of a large number of boride compositions synthesized over a wide range of concentrations of the original materials, we found that samples obtained with an excess of the metal decomposed more or less energetically on interacting with dilute HCl, with the formation of boranes. This indicates the presence of as-yet unknown lower borides, differing in composition and properties from MeB_4 and MeB_6, in the cakes.

The experimental data obtained by studying the chemical properties of boride compositions involving alkaline-earth and rare-earth metals enabled us to evolve a method for the chemical phase analysis of these (scheme on p. 107).

According to this method, the interaction products derived from the metals and boron are sifted and weighed and then treated with very dilute HCl (1:10) in the cold until decomposition ceases. The insoluble residue is filtered off and weighed. The filtrate is subjected to chemical analysis for the boron and metal content by ordinary analytical methods.

Chemical Phase Analysis of Products Formed in the Me–B System

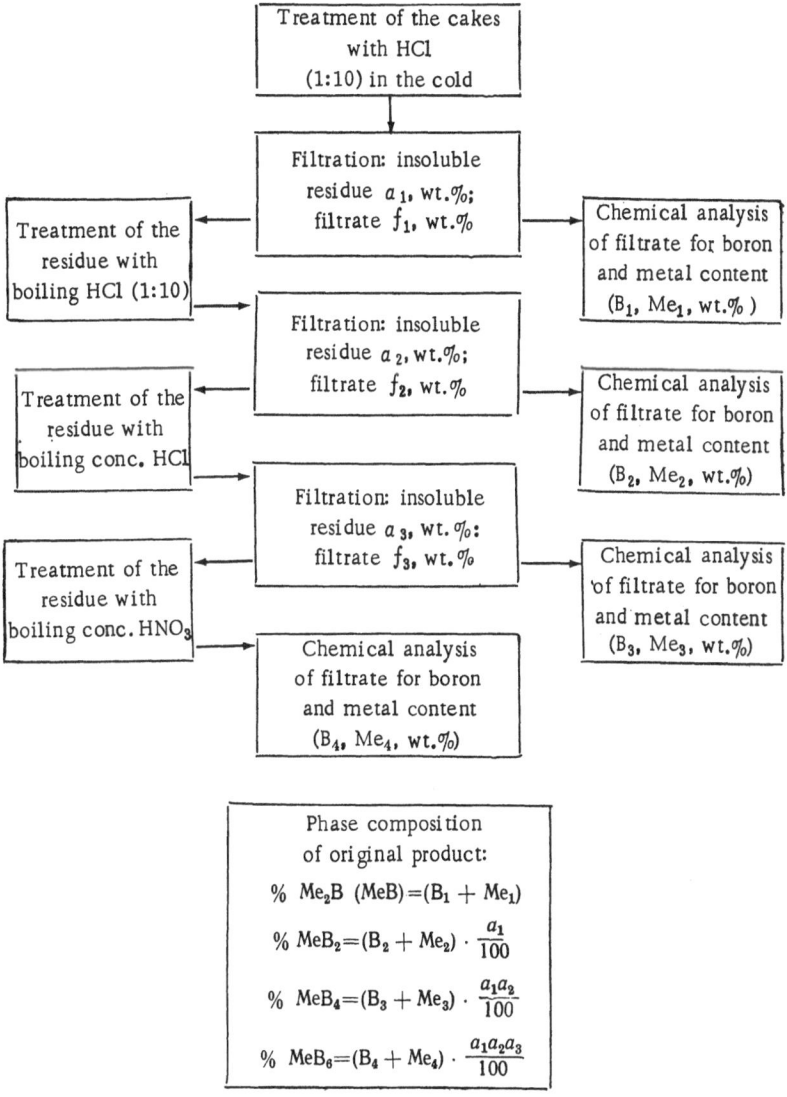

The resultant experimental data showed that the sintered masses formed by lanthanum and gadolinium with boron were decomposed most rapidly and completely in dilute HCl. The data presented in Table 1 show that in this case the lanthanum and boron or gadolinium and boron pass into the solution in quantities corresponding to atomic ratios 2:1 and 1:1 respectively. This shows that the La–B and Gd–B systems form readily decomposed boride phases of compositions La_2B and GdB.

The residues of the boride cakes not dissolving in dilute HCl at room temperature are then treated with boiling HCl (1:10). The experiments showed that as a result of this treatment the samples obtained by the sintering of calcium and gadolinium with boron decomposed energetically with the formation of boranes. According to the chemical analysis, the ratio between the metal and boron passing into solution is close to 1:2. We may therefore conclude that calcium and gadolinium form boride phases of composition MeB_2.

The cakes of cerium and boron decomposed very little on interacting with dilute HCl and apparently contained no easily hydrolyzed lower borides.

TABLE 1. Results of the Hydrolytic Decomposition of the Interaction
with Boron, Obtained under the Optimum Conditions

Atomic composition of reacting mixture	Interaction temperature, °C	Passes into filtrate on decomposition of borides in HCl (1:10), %							
		at room temperature				at boiling point			
		Me	B	com-bined	atomic ratio Me:B	Me	B	com-bined	atomic ratio Me:B
Ca + 2B	900	—	—	—	—	28.0	17.1	45.1	1 : 2.21
Ca + 4B	1000	—	—	—	—	48.1	26.9	75.0	1 : 2.1
Ca + 6B	1300	—	—	—	—	—	—	—	—
La + 2B	900	86.5	3.4	89.9	2 : 1	—	—	—	—
La + 2B	1000	46.4	2.0	48.4	1.9 : 1	—	—	—	—
La + 4B	1000	33.6	1.5	35.1	2 : 1	—	—	—	—
La + 6B	1300	—	—	—	—	—	—	—	—
Gd + B	1200	87.5	5.8	93.3	1 : 0.9	—	—	—	—
Gd + 2B	1300	—	—	—	—	83.5	11.1	94.6	1 : 1.9
Gd + 4B	1400	—	—	—	—	—	—	—	—
Gd + 6B	1400	—	—	—	—	—	—	—	—

The products not decomposed in HCl (1:10) are further subjected to prolonged boiling with concentrated HCl. The resultant solid residues are filtered off, weighed, and boiled with nitric acid. The HCl and HNO_3 solutions are subjected to chemical analysis for metal and boron content.

According to the data presented in Table 1, the tetrabromides of the rare-earth metals dissolve in concentrated hydrochloric acid, and the hexaborides of these metals and also that of calcium in nitric acid. Clearly there is no calcium tetraboride. The existence of the lower boride phase of calcium, lanthanum, and gadolinium is also confirmed by x-ray analysis, the corresponding lines being observed on the powder diffraction photographs of these borides.

Chemical Phase Analysis of a Number of Ternary

Me−B−C Systems

It was found in the course of the investigations cited earlier that ternary phases were formed when certain metals interacted with a mixture of boron and carbon. Thus there are borocarbides of alkaline-earth [3] and rare-earth [4, 6] metals, beryllium [7], chromium, and manganese [11]. Hence the Me−B−C systems form borocarbides (carboborides) as well as the carbides and borides of the metals and boron carbide.

In order to separate these compounds and determine their composition and their proportions in test samples, the method of chemical phase analysis constitutes an efficient procedure. There are two general schemes of analysis for application to ternary systems. One of these is used for cases in which the metal forms easily hydrolyzed borocarbides with boron and carbon, for example, in determining the borocarbides of the rare-earth metals, and the second is used in the chemical phase analysis of products containing chemically stable borocarbides, for example, those of beryllium.

For the chemical phase analysis of cakes obtained by the interaction of alkaline-earth metals and certain lanthanides (La, Ce, Gd, Eu) with boron and carbon, the following sequence of operations is proposed.

1. A chemical analysis of the original product is carried out for its metal, boron, and carbon content (Me_1, B_1, C_1).

2. Using individual samples, the presence of carbides of the MeC_2 type is determined, together with the corresponding metal and carbon concentration (Me_5, C_5); the determination

Products of Certain Alkaline-Earth and Rare-Earth Metals
for the Synthesis of Individual Boride Phases

Passes into filtrate on decomposition of borides by boiling in acids, %							
HCl conc.				HNO$_3$ conc.			
Me	B	com-bined	atomic ratio Me:B	Me	B	com-bined	atomic ratio Me:B
—	—	—	—	—	54.0	54.0	—
—	—	—	—	—	24.2	24.2	—
—	—	—	—	37.7	61.5	99.2	1:6.0
—	—	—	—	—	9.8	9.8	—
39.0	12.6	51.6	1:4.1	—	—	—	—
44.9	14.0	58.9	1:4.0	9.0	4.2	13.2	1:6.0
—	—	—	—	67.0	32.2	99.2	1:6.1
—	—	—	—	—	—	—	—
78.2	21.4	99.6	1:3.9	—	—	—	—
—	—	—	—	69.8	29.8	99.6	1:6.0

is based on the amount of acetylene evolving when the cakes are treated with water. It should be noted that usually there are no such carbides in the cakes, since if any are produced they interact with boron forming borocarbides.

3. The original cake is treated with dilute HCl. The borocarbides and carbides of the metals then generally decompose.

The lower borides of the alkaline-earth and rare-earth metals are unstable in the presence of carbon and are never found in the cakes under consideration.

As shown earlier [12], the hydrolysis of borocarbides in acid solutions takes place very energetically and is accompanied by the formation of solid, liquid, and gaseous (not containing acetylene) organic products.

4. The residue not dissolving in HCl is filtered and washed first with water and then with acetone, which dissolves solid organic products of the hydrolysis of the borocarbides.

5. The filtrate is analyzed for boron and metal content (B_2, Me_2).

No allowance is made in the proposed scheme for the small quantity of boron remaining in the organic substances. This simplification of the analysis has no serious effect on the determination of the composition of the borocarbides.

6. We calculate the amount of carbon transferred to the organic substances in the hydrolysis of the borides and borocarbides (C_2) as the difference between the content of the cakes before and after hydrolysis.

7. On the basis of the proportions of metal, boron, and carbon determined (B_2, Me_2, C_2) (after subtracting the carbide metal and the carbon) the composition of the borocarbide subject to hydrolytic decomposition is calculated.

8. The residue not dissolving in HCl is treated with nitric acid; the hexaboride then decomposes, and B_4C, graphite, or a mixture of these remains in the residue.

9. The nitric acid filtrate is analyzed for boron and metal content (B_3, Me_3), the total amount of this corresponding to the amount of MeB_6 in the cake.

10. In the residue left after the nitric acid treatment we determine the boron and calculate the corresponding amount of B_4C.

Chemical Phase Analysis of Products Formed by the Interaction
of Alkaline-Earth and Rare-Earth Metals with Boron and Carbon

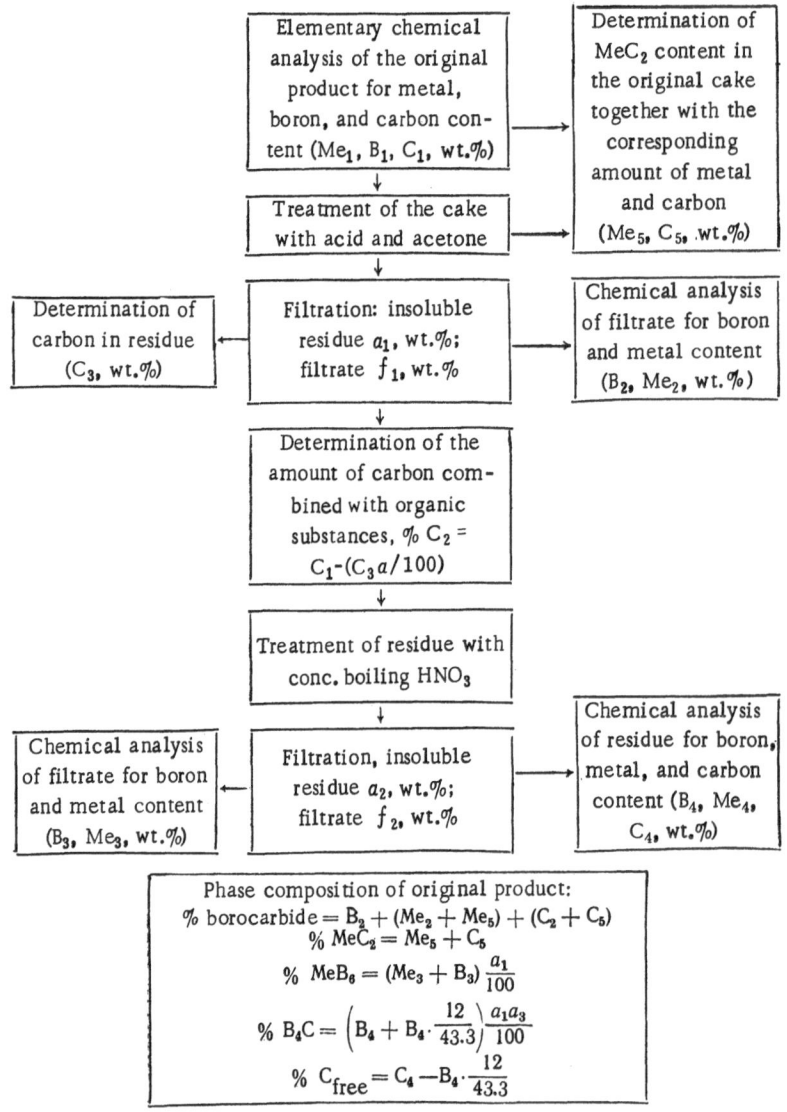

TABLE 2. Results of the Hydrolytic Decomposition
of Products Formed by the Interaction of Calcium
and Lanthanum with a Mixture of Boron and
Carbon, Obtained under the Optimum Conditions
for the Synthesis of the Borocarbides

Atomic composition of the reacting mixture	Interaction temperature, °C	Decomposed in the treatment of the cakes with HCl (1:4), %				Atomic ratio Me:B:C (from hydrolytic-decomposition data)
		Me	B	C	combined	
Ca + 2B + 2C	1300	35.8	10.2	21.8	67.8	1 : 2 : 1
Ca + 2B + 4C	1900	35.0	21.1	41.8	97.9	1 : 2 : 4
La + B + C	1300	80.5	6.6	7.6	94.7	1 : 1 : 1.1
La + B + 2C	1900	78.5	7.2	12.1	97.8	1 : 1.2 : 1.8
La + 2B + 4C	1900	64.0	9.5	22.9	96.4	1 : 1.9 : 4.1

Chemical Phase Analysis of Products Containing Chemically Stable Borocarbides

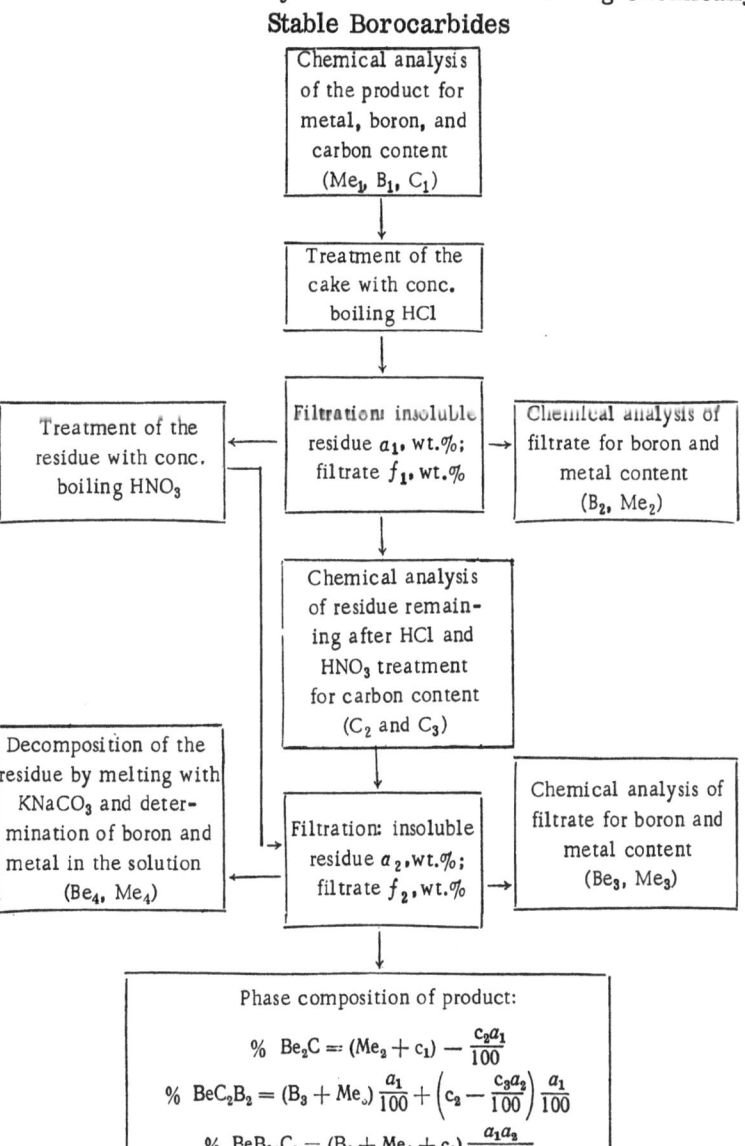

11. In conclusion, after analyzing all the solid residues and acid solutions chemically, the phase composition of the original product is calculated from the formulas presented in the scheme of p. 110.

By way of example, Table 2 shows the results of the hydrolytic decomposition and chemical phase analysis of the products formed by the interaction of calcium and lanthanum with boron and carbon. According to these data, the system Ca–B–C contains two borocarbide phases of composition CaC_2B and CaC_4B_2, and the La–B–C system contains three: $LaBC$, LaC_2B, and LaC_4B_2.

In the same way we were able to identify borocarbides of strontium and barium of the same composition as those of calcium, and also borocarbides of cerium, neodymium, praseodymium, and gadolinium.

In studying the products formed in the Me–B–C system we also carried out an x-ray phase analysis of the original materials and the residues not dissolving in acids. The results

TABLE 3. Results of the Acid Decomposition of Products Formed by the Interaction of Beryllium with Boron and Carbon at 2000°C

Molar composition of reacting mixture Be:C:B	Subject to decomposition in the boiling of the cakes									Chemical analysis of the products not dissolved in HCl and HNO3				
	HCl				HNO₃									
	Be	C	combined, %	atomic ratio, Be:C	Be	B	C	combined, %	atomic ratio, Be:B:C	Be	B	C	combined, %	atomic ratio, Be:B:C
2:1:1	12.3	8.9	21.2	1.8:1	12.8	30.0	30.9	73.7	1:1,8:2	—	—	—	—	—
1:1:1	30,2	24.8	55.0	1.9:1	5.8	14.2	15.0	35.0	1:2:2	—	—	—	—	—
1:2:2	—	—	—	—	16.6	38.2	43,6	98.4	1:1,9:2	—	—	—	—	—
1:2:4	—	—	—	—	6.2	15.2	16.5	37.9	1:2:2	3,5	48.5	10.2	62.2	1:11.7:2
1:2:12	—	—	—	—	—	—	—	—	—	5,3	79.1	14,1	98,5	1:12:2

confirmed our chemical phase analysis, in particular the existence of all the borocarbide phases presented in Table 2.

The chemical phase analysis of sintered masses containing chemically-resistant borocarbides (such as the Be−B−C system, see p. 111) comprises the following steps.

1. The test sample is subjected to chemical analysis for boron, metal, and carbon content (B_1, Me_1, C_1).

2. The original cake is treated with concentrated, boiling HCl.

3. The residue not dissolving in HCl is filtered off.

4. The filtrate is analyzed for boron and metal content (B_2, Me_2).

5. The product not dissolving in HCl is further boiled in HNO_3. The solid residue is filtered off.

6. The nitric acid filtrate is analyzed for beryllium and boron content (Me_3, B_3).

7. The amount of carbon corresponding to the C concentration in the compounds decomposed by the HCl and HNO_3 is calculated (from the difference between the carbon contents of the product before and after each acid treatment).

8. The solid residue obtained after treating the samples with nitric acid is decomposed by melting in $KNaCO_3$ and then dissolving the cake in acid. Beryllium and boron are determined in the solution. The amount of carbon in the residue is determined from another sample.

9. Using the chemical-analysis data, the phase composition of the material in question is calculated.

According to the experimental data presented in Table 3, when the original cakes interact with concentrated HCl a compound containing only beryllium and carbon is decomposed ($B_2 = 0$). The Be:C ratio is close to 2:1, i.e., it corresponds to a carbide Be_2C. Hence the reaction between boron, beryllium, and carbon produces no easily hydrolyzed borocarbides.

According to the chemical analysis, nitric acid decomposes a ternary compound containing beryllium, boron, and carbon in 1:2:2 ratio, i.e., a borocarbide of composition BeC_2B_2. With increasing amount of boron in the original mixture, cakes principally containing a compound dissolving in neither HCl nor HNO_3 are formed. This substance, after freeing from traces of B_4C and BeC_2B_2 by acid treatment, contains beryllium, carbon, and boron in the atomic ratio 1:2:12, and thus constitutes a beryllium borocarbide of composition BeC_2B_{12}.

The existence of two borocarbide phases of beryllium was also confirmed by x-ray phase analysis. By the example of the binary and ternary systems considered, we see that chemical phase analysis, supplemented by x-ray investigations, yields completely reliable conclusions as to the phase composition of the cakes under examination and the proportion of individual components in these. This method also offers the possibility of identifying new binary and ternary compounds and determining their composition.

Conclusions

1. In determining the phase composition of products formed in certain boride and borocarbide systems, excellent results are obtained by chemical phase analysis based on the different chemical stability of the compounds formed in the Me–B and Me–B–C systems during hydrolytic decomposition and interaction with acids.

2. By way of an example, we have considered the chemical phase analysis of boride compositions of alkaline- and rare-earth elements. We have demonstrated the existence of new lower boride phases of these metals and determined their composition.

3. We have considered two general schemes for the chemical phase analysis of products formed in certain Me–B–C systems. One of these may be used when the metal forms easily hydrolyzed borocarbides with boron and carbon, for example, the borocarbides of the alkaline-earth metals and the lanthanides; the second may be used for the chemical phase analysis of products containing chemically stable borocarbides, for example, those of beryllium.

4. By way of example, we have presented the results of a chemical phase analysis of the products formed by the interaction of lanthanum, calcium, and beryllium with boron and carbon; these confirm the existence of ternary phases in these systems.

5. The conclusions derived from chemical phase analysis should always be confirmed by x-ray phase analysis.

Literature Cited

1. L. Ya. Markovskii and N. V. Vekshina, Zh. Priklad. Khim., 34:16 (1961).
2. L. Ya. Markovskii and N. V. Vekshina, Zh. Priklad. Khim., 38:1945 (1965).
3. L. Ya. Markovskii and N. V. Vekshina, Zh. Priklad. Khim., 34:342 (1961).
4. L. Ya. Markovskii, N. V. Vekshina, and G. F. Pron', Zh. Priklad. Khim., 35:2090 (1962).
5. L. Ya. Markovskii and N. V. Vekshina, Zh. Priklad. Khim., 37:2126 (1964).
6. L. Ya. Markovskii, N. V. Vekshina, and G. F. Pron', Zh. Priklad. Khim., 38:245 (1965).
7. L. Ya. Markovskii et al., Zh. Priklad. Khim., 39:13 (1966).
8. L. Ya. Markovskii and Yu. D. Kondrashev, Zh. Priklad. Khim., 2:34 (1957).
9. L. Ya. Markovskii, G. V. Kaputovskaya, and Yu. D. Kondrashev, Zh. Priklad. Khim., 4:1710 (1959).
10. L. Ya. Markovskii and E. T. Bezruk, Zh. Priklad. Khim., 35:491 (1962).
11. L. Ya. Markovskii and E. T. Bezruk, Zh. Priklad. Khim., 34:258 (1956).
12. L. Ya. Markovskii and N. V. Vekshina, 37:2120 (1964).

INTERACTION OF BORIDES WITH
CARBON AND CARBIDES

L. Ya. Markovskii, N. V. Vekshina, and E. T. Bezruk

State Institute of Applied Chemistry

The question as to the interaction of borides with carbon and carbides is of great scientific and practical interest. The practical importance is associated with the fact that many borides find applications in contact with carbon and carbides, which are often used as construction materials.

A study of the reaction of borides with carbon and carbide will supplement our knowledge as to the chemical properties of these compounds and the transformations involved in their interaction with carbon and carbides at high temperatures. The problem has been relatively little studied. There is only a paper by Glaser [12] regarding the interaction of the borides of the transition metals with carbon, in which it is shown that the borides have a high stability with respect to the latter, and articles by Benesovsky et al. [14, 15] devoted to a study of ternary systems incorporating Group IV, V, and VI elements and actinides together with carbon, as well as several papers by G. V. Samsonov et al. relating to solid-phase reactions of refractory compounds. We should also mention the investigations of the State Institute of Applied Chemistry into ternary Me−B−C systems, in which the existence of borocarbides of Group II and III metals and also of chromium and manganese has been confirmed [3-6]. The formation of these compounds must be taken into account when considering transformations associated with the contact of borides with carbon-containing substances.

Interaction of the Borides of Group II and III Metals

with Carbon and Carbides

Among the Group II metals, magnesium forms a number of borides MgB_2, MgB_6, MgB_{12} [9], which decompose after roasting with carbon at high temperatures (2000°C). The magnesium evaporates, while the boron interacts with the carbon and forms B_4C. The magnesium borides form no ternary compounds with carbon. For the alkaline-earth metals only one type of boride phase is well known, namely, the hexaborides of composition CaB_6, SrB_6, and BaB_6. The interaction of the hexaborides with carbon was studied by roasting mixtures of the boride and graphite at 1300 and 2000°C [1].

The resultant samples were subjected to chemical and x-ray phase analysis. The chemical analysis showed that no interaction took place with water or HCl and no trace of acetylene or boron-organic substances appeared, such as would occur if carbides or borocarbides of the metals were formed.

TABLE 1. Results of the Interaction
of Beryllium Borides with Carbon
and Carbides (t = 2000°C)

Molar composition of reaction mixture Be:C:B	Chemical phase analysis data for the cakes, %				X-ray data
	BeC_2B_2	$BeB_{12}C_2$	B_4C	combined	
$BeB_2 + 2C$	97.0	—	3.2	100.2	Lines of BeC_2B_2
$BeB_6 + 2C$	23.3	74.0	—	97.0	Lines of BeC_2B_2 and $BeB_{12}C_2$
$BeB_{12} + 2C$	—	99.9	—	99.9	Lines of $BeB_{12}C_2$
$BeB_2 + B_4C +$					
$+ B_2C + 4C$	98.0	—	—	98.0	Lines of BeC_2B_2
$BeB_6 + Be_2C + 5C$	96.0	—	—	96.0	
$BeB_6 + 1,5B_4C + C$	—	98.0	—	98.0	Lines of $BeB_{12}C_2$
$BeB_{12} + Be_2C + 5C$	43.0	57.0	—	100	Lines of BeC_2B_2 and $BeB_{12}C_2$

The powder diffraction patterns of the products contained the lines of MeB_6 and graphite. On boiling in HNO_3 the hexaboride dissolves and only graphite remains in the residue. Thus the Me_6–graphite mixture is unaltered by roasting. Hence the hexaborides of the alkaline-earth metals fail to react with carbon.

The high chemical resistance of the hexaborides with respect to carbon is also supported by the formation of borides MeB_6 as a result of the interaction of carbides of composition MeC_2 with boron, and also as a result of the reduction of the corresponding oxides by boron carbide.

In earlier results obtained in the State Institute of Applied Chemistry [7], the existence of diborides of the alkaline-earth metals was established; in contrast to the hexaborides, these interact fairly easily with carbon (at 1300°C), forming ternary phases, namely, borocarbides of composition MeC_2B.

Beryllium gives many boride phases, including the borides Be_4B, BeB_2, BeB_2, BeB_4, BeB_6, BeB_{12} [2, 10, 11]. All these decompose on roasting in a mixture with carbon. In addition to beryllium carbide, borocarbides of beryllium are formed: BeC_2B_2 (hexagonal structure) or $BeB_{12}C_2$ (B_4C type of structure).

At high temperatures the beryllium borides interact not only with carbon but also with beryllium and boron carbides, also yielding borocarbides distinguished by high chemical stability (Table 1).

Of the borides of Group III metals, the interaction of a number of hexaborides of the rare-earth metals, including lanthanum, cerium, gadolinium, and europium, with carbon has been studied.

TABLE 2. Composition of the Products Formed by the
Interaction of Chromium Borides with Carbon
and Carbides (t = 1600°C)

Molar composition of reaction mixture	Chemical phase analysis data for the cakes, %						X-ray analysis data
	CrB	CrB_2	Cr_7BC_4	$Cr_3C_2 + B$ (up to 10 at.%)	C	combined	
$Cr_2B + C$	46.1	—	—	53,9	—	100	Lines of CrB and Cr_3C_2
$CrB + C$	82.8	—	—	—	17,1	99,9	Lines of CrB and C
$CrB_2 + C$	—	85,0	—	—	15,0	100	Lines of CrB_2 and C
$Cr_2B + Cr_3C_2$	—	—	—	100	—	100	Lines of Cr_3C_2
$CrB + Cr_3C_2$	—	—	—	100	—	100	Lines of Cr_3C_2
$CrB_2 + Cr_3C_2$	44.8	—	—	54.5	—	99.3	Lines of CrB and Cr_3C_2
$Cr_2B + Cr_7C_3$	—	—	87.1	12.7	—	99.8	Lines of Cr_7BC_4
$CrB + Cr_7C_3$	—	—	81.1	18.9	—	100	Lines of Cr_7BC_4
$CrB_2 + Cr_7C_3$	—	—	—	100	—	100	Lines of Cr_3C_2

TABLE 3. Results of Experiments Relating to the Interaction of Manganese Borides with Manganese Carbides and Boron Carbide

Composition of original mixture molar ratio	Roasting temperature, °C	Phase composition of interaction products derived from chemical and x-ray phase analysis
$Mn_4B + Mn_{23}C_6$	1300	Mn_8BC
$MnB + Mn_7C_3$	1300	$Mn_7BC_2 + Mn_8BC$
$Mn_2B + Mn_{23}C_6$	1300	Mn_8BC
$Mn_2B + Mn_{23}C_6$	1600	Mn_7BC_2
$Mn_2B + Mn_3C$	1800	Mn_7BC_2
$MnB + Mn_7C_3$	2000	$Mn_7BC_2 + C$
$MnB + B_4C$	2000	$MnB + Mn_7BC_2$

The experimental results showed that the hexaborides of the rare-earth elements, like those of the alkaline-earth metals failed to interact with carbon. On the other hand, the recently obtained lower borides of lanthanum [8] and gadolinium interact with carbon to form borocarbides of various compositions, including MeC_2B, MeC_4B_2, and $MeCB$.

Interaction of Chromium Borides with Carbon and Carbides

Table 2 presents the results of chemical and x-ray phase analyses of products obtained as a result of the sintering of chromium borides Cr_2B, CrB, and CrB_2 with carbon and chromium carbide at 1600°C.

It follows from these data that the different boride phases of chromium behave differently with respect to carbon and carbides. The chromium mono- and diboride fail to interact with carbon. The lower boride, Cr_2B, is unstable in the presence of carbon and transforms into CrB and Cr_3C_2. On baking the chromium borides with carbides, we obtain products with powder photographs showing no boride lines; hence the borides interact with the carbides. According to the results of chemical and x-ray phase analyses, either solid solutions of borides in the Cr_3C_2 lattice or else a ternary compound, chromium borocarbide (Cr_7BC_4) is formed.

Interaction of Manganese Borides with Carbon and Manganese Carbides [3]

Of the manganese borides, only the monoboride is stable in the presence of carbon at high temperatures. In the presence of carbon the borides Mn_3B_4 and MnB_2 transform into a boride of composition MnB, while the lower borides form borocarbides with carbon.

On heating to 1300°C, alloys of composition Mn_4B interact with carbon to form a mixture of borocarbides Mn_7BC_2 and Mn_8BC, although for temperatures in excess of 1500°C only the former of these appears.

It follows from the experimental data obtained (Table 3) that, when manganese borides interact with carbides, borocarbides of composition Mn_8BC (at 1300°C) and Mn_7BC_2 (at higher temperatures) are obtained.

The manganese borocarbide of composition Mn_7BC_2 possesses ferromagnetic properties and has the structure of the hexagonal carbide Cr_7C_3, while the borocarbide Mn_8BC is nonferromagnetic and has the structure of the complex cubic carbide $Cr_{23}C_6$. Both these are hydrolytically unstable. On decomposition by water and acids, liquid and gaseous organic compounds are evolved.

The fact that the manganese borocarbides obtained have a structure similar to the carbides Mn_7C_3 and $Mn_{23}C_6$ clearly shows that they constitute the derivatives of the carbides and are formed as a result of the replacement of carbon atoms by boron in the corresponding crystal lattices.

Discussion of Results

Table 4 gives a summary of existing experimental and published data relating to the interaction of borides with carbon. We see that, depending on their composition, structure, and chemical properties, the borides behave in a variety of different ways toward carbon.

TABLE 4. Stability of Borides with Respect to Carbon

Metal	Me_4B	Me_2B	Me_3B_2	MeB	Me_3B_4	MeB_2	Me_2B_5	MeB_4	MeB_6	MeB_{12}
Me	U*	—	U*	—	—	U*	—	U*	U*	U*
Mg	—	—	—	—	—	U	—	U	—	U
Ca, Sr, Ba	—	—	—	—	—	U*	—	—	S	—
Lanthanides	—	—	—	—	—	U	—	—	S	—
Ti*	—	U	—	U	—	S	—	—	—	—
Zr*	—	—	—	U	—	S	—	—	—	U
Hf*	—	—	—	S	—	S	—	—	—	—
Th*	—	—	—	—	—	—	—	U*	S	—
V, Nb*, Ta	—	—	U	U	U	S	—	—	—	—
Cr	—	U*	—	U	—	S	—	—	—	—
Mo*	—	U*	—	S	—	S	U	—	—	U
W*	—	U	—	S	—	S	S	—	—	U
Mn	—	U*	—	S	U	S	—	—	—	—

Note. U = unstable in presence of carbon; U* = unstable in presence of carbon, forms borocarbide; S = stable in presence of carbon; * signifies published data [12-15].

All the lower borides of the metals of the Me_4B, Me_2B_4, Me_3B_2 type, i.e., boride phases with a large metal content, are unstable with respect to carbon and interact with it, forming either carbides or borocarbides.

As we increase the amount of boron in the borides, thus strengthening the boron–boron bond, the stability increases. Thus, in the hexaborides, which have a skeletal structure of boron atoms, the stability is greater than in the diborides, in the latter greater than in the monoborides, and so on. This general law clearly appears in other chemical reactions of the borides, in particular hydrolytic decomposition, and in general features it is also applicable to reactions between the borides and carbon or carbides. However, in this case the matter is a little more complicated, and there are some exceptions as yet not quite understood. Thus for all the dodecaborides studied we find instability with respect to carbon. Among the manganese borides the most stable phase with respect to carbon is the monoboride. It should be noted that a number of other manganese borides, for example, Mn_3B_4 and MnB_2, rapidly change into MnB in the presence of carbon. In addition to this, since the borides interact with carbon at high temperatures, the energy of the boron–boron bond is not always the decisive factor, and the relation between the affinity of the metal toward carbon and boron respectively plays a major part.

We may suppose that a further study of the mutual transformations characterizing the reactions between borides and carbides will reveal the main laws governing these.

Literature Cited

1. N. V. Vekshina and L. Ya. Markovskii, Zh. Priklad. Khim., 34:2171 (1961).
2. G. S. Markevich and L. Ya. Markovskii, Trudy Gos. Inst. Priklad. Khim., 45:139 (1960).
3. L. Ya. Markovskii and E. T. Bezruk, Zh. Priklad. Khim., 39:258 (1966).
4. L. Ya. Markovskii and N. V. Vekshina, Zh. Priklad. Khim., 34:242 (1961).
5. L. Ya. Markovskii et al., Zh. Priklad. Khim., 13:39 (1966).
6. L. Ya. Markovskii, N. V. Vekshina, and G. F. Pron', Zh. Priklad. Khim., 35:2091 (1962).
7. L. Ya. Markovskii and N. V. Vekshina, Zh. Priklad. Khim., 34:16 (1961).
8. L. Ya. Markovskii and N. V. Vekshina, Zh. Priklad. Khim., 38:1945 (1965).

9. L. Ya. Markovskii, Yu. D. Kondrashev, and G. V. Kaputovskaya, Zh. Organ. Khim., 25:443 (1955).

10. L. Ya. Markovskii, Yu. D. Kondrashev, and G. V. Kaputovskaya, Zh. Organ. Khim., 25:1045 (1

11. L. Ya. Markovskii and G. S. Markevich, Zh. Priklad. Khim., 33:1667 (1960).

12. F. Glaser, J. Metals, 4:391 (1952).

13. H. Nowotny, E. Rudy, and F. Benesovsky, Mn. Chem., 92:393 (1961).

14. E. Rudy, F. Benesovsky, and L. Toth, Z. Metallkunde, 54(6):345 (1963).

15. L. Toth, H. Nowotny, F. Benesovsky, and E. Rudy, Mn. Chem., 92:794, 956 (1961).

CHEMICAL PROPERTIES AND ANALYSIS OF CERTAIN TRANSITION METAL SULFIDES

S. V. Radzikovskaya and V. F. Bukhanevich

Institute of Problems in Materials Sciences, Academy of Sciences of the Ukrainian SSR

The sulfides of transition metals belong to a group of refractory materials little studied as regards methods of production and physicochemical properties. In the transition metal-sulfur systems, we only know the compositions of the phases obtained by synthesis from the elements and a few crystal-chemical properties (types of structure, lattice constants). There is no published information regarding the chemical stability of the sulfides in various media nor their oxidation resistance. In view of this it would appear interesting to study the chemical properties of transition metal sulfides and on the basis of these develop methods for their analysis.

We studied various sulfide phases of a number of transition metals: Sc_2S_3, HfS_2, V_2S_3, $NbS_{1.6}$, TaS_2, Cr_2S_3. These were obtained by the action of H_2S on the powdered metals or their oxides at 1300-1500°C [1, 2, 3]. The crystal-chemical properties and chemical compositions of the sulfides in question are given in Table 1.

We studied the stability of the sulfide powders in solutions of various reagents on heating for 1 h. The sulfide sample was placed in a conical flask furnished with a reflux condenser,

TABLE 1. Crystal-Chemical Properties of the Sulfides of Certain Transition Metals

Sulfide	Color	Structure	Lattice constants			Density, g/cm³		Chemical composition, %	
			a	b	c	pyknometric	x-ray diffraction	Me_{tot}	S_{tot}
Sc_2S_3	Yellow-orange	Ortho-rhombic	5.18	—	—	—	2.96	45.8	55.1
HfS_2	Brownish-red	Trigonal	3.63	—	5.85	—	—	73.4	25.3
V_2S_3	Black	Hexagonal	6.59	—	22.13	—	—	48.5	51.0
$NbS_{1.6}$	The same	Hexagonal	3.338	—	17.827	5.9	6.0	62.2	35.4
TaS_2	Black with a green tinge	Trigonal	3.36	—	5.89	7.1	7.16	73.0	25.5
Cr_2S_3	Black	The same	5.942	—	11.125	3.6	—	52.0	48.6

119

TABLE 2

Sulfide	Reagent	Insoluble residue	Chemical composition of insoluble residue, %		Amount of metal in solution, %	Notes
			S_{tot}	Me_{tot}		
Sc_2S_3	H_2SO_4, $d = 1.84$	0	—	—	—	Complete decomposition
HfS_2		0	—	—	0	
V_2S_3		0	—	—	51.2	
TaS_2		0	—	—	73.0	
Cr_2S_3		0	—	—	—	
Sc_2S_3	H_2SO_4, 1 : 1	0	—	—	—	—
HfS_2		0	—	—	—	
V_2S_3		59.5	52.4	39.8	25.7	
TaS_2		99.6	22.9	73.0	1.8	
Cr_2S_3		99.3	—	—	0.25	
Sc_2S_3	HCl, $d = 1.19$	0	—	—	—	—
HfS_2		0	—	—	—	
V_2S_3		96.0	48.2	51.4	1.7	
TaS_2		98.0	24.0	73.0	1.1	
Cr_2S_3		100.1	0	0	0	
Sc_2S_3	HCl = 1 : 1	0	—	—	—	—
HfS_2		0	—	—	—	
V_2S_3		96.3	48.1	51.0	1.7	
TaS_2		98.7	23.2	72.9	1.0	
Cr_2S_3		99.9	0	0	0	
Sc_2S_3	HNO_3, $d = 1.4$	0	—	—	—	Complete decomposition
HfS_2		0	—	—	—	
V_2S_3		0	—	—	51.2	
TaS_2		0	—	—	73.0	
Cr_2S_3		0	—	—	52.0	
Sc_2S_3	HNO_3, 1 : 1	0	—	—	—	Complete decomposition
HfS_2		0	—	—	—	
V_2S_3		0	—	—	51.2	
TaS_2		0	—	—	73.0	
Cr_2S_3		0	—	—	52.0	
Sc_2S_3	H_3PO_4, $d = 1.21$	—	—	—	—	—
HfS_2		—	—	—	—	
V_2S_3		—	—	—	—	
TaS_2		1.4	22.0	72.9	51.0	
Cr_2S_3		99.9	0	0	0	
Sc_2S_3	H_2O_2, 30%	0	—	—	—	Complete decomposition
HfS_2		0	—	—	—	
V_2S_3		0	—	—	19.7	
TaS_2		0	—	—	73.0	
Cr_2S_3		92.1	—	—	3.7	
HfS_2	$H_2C_2O_4$, 6%	93.0	48.4	51.0	3.8	—
V_2S_3		79.6	25.0	73.0	14.8	
TaS_2		99.9	0	0	0	
Sc_2S_3						
Cr_2S_3						
Sc_2S_3	Bromine water	0	—	—	—	—
HfS_2		0	—	—	—	
V_2S_3		74.8	51.2	47.2	15.0	
TaS_2		—	—	—	—	
Cr_2S_3		81.2	—	—	7.75	

TABLE 2. (continued)

| Sulfide | Reagent | Insoluble residue | Chemical composition of insoluble residue, % | | Amount of metal in solution, % | Notes |
			S_{tot}	Me_{tot}		
Sc_2S_3	H$_2$O	99,5	—	—	—	—
HfS_2		82.4	—	—		
V_2S_3		99,4	48,4	51.2	0	
TaS_2		99.8	48.4	73.0	0	
Cr_2S_3		100.0	0	0	0	
HfS_2	NaOH, 40%	0	—	—	—	—
V_2S_3		93.5	47.7	50.5	5.5	
TaS_2		97.0	6.0	72.8	2.4	
Cr_2S_3		98.0	0	0	0	
Sc_2S_3	NaOH, 10%	0	—	—	—	—
V_2S_3		98.6	47,6	55.2	0	
TaS_2		36.0	9.3	72.9	55.5	
Cr_2S_3		99.5	0	0	0	
HfS_2		97.9	—	—	—	

TABLE 3. Oxidation of Niobium and Tantalum Sulfides in Oxygen

Temp., °C	Amount of oxidized sulfur, wt. %															
	$NbS_{1.6}$								TaS_2							
	Oxidation time, min															
	10	20	30	40	50	60	90	120	10	20	30	40	50	60	90	120
300	0.7	1.3	2.1	2.2	2.9	3.2	4.5	5,4	1.4	1,4	2.5	2,5	2.5	2.5	2.7	4.0
350	2.3	11.2	14.5	15.9	18.4	21.2	25.9	29.0			Not studied					

N o t e. The complete oxidation of NbS$_{1.6}$ at 400°C and TaS$_2$ at 500°C occurs in 10 min; for TaS$_2$ the amount of oxidized sulfur at 400-450°C after 2 h was 19.9-20.8 .

and a solution of reagent was added. The insoluble residue was filtered through a Schott crucible, and the amount of sulfur and metal in the residue was determined. The metal passing into solution as a result of the decomposition of the sulfide was determined in the filtrate. The stabilities of the Sc$_2$S$_3$, HfS$_2$, V$_2$S$_3$, TaS$_2$, Cr$_2$S$_3$ powders so determined are presented in Table 2.

The powders of all the sulfides studied are stable in air at room temperature for long periods and also in boiling water. The sulfides Sc$_2$S$_3$ and Cr$_2$S$_3$ even remain undecomposed on boiling in alkali solutions. All the sulfides are completely dissolved by oxidizing acids (nitric, dilute sulfuric) and by bromine and hydrogen peroxide solutions.

We also studied the oxidizability of the sulfides on heating in a current of oxygen. The powder particle size was 270 mesh and the velocity of the current of oxygen 200 ml/min. The degree of oxidation was estimated from the amount of SO$_2$ evolved.

Table 3 presents the results of the oxidation of niobium and tantalum sulfides. We see from these results that NbS$_{1.6}$ and TaS$_2$ start oxidizing at 300°C, complete oxidation setting in at 400-500°C. Analogous results were obtained for the sesquisulfides of scandium and vanadium and hafnium disulfide.

Chromium sesquisulfide Cr_2S_3 is oxidized at a higher temperature, 500°C, complete oxidation setting in at 1000°C. The final oxidation products are the oxides of the metals.

Thus the sulfides of scandium, hafnium, vanadium, niobium, and tantalum, like other refractory sulfides, are unstable on heating in air. The easy oxidizability of the sulfides on heating in air offers the possibility of determining the total metal content by roasting a sulfide sample, simultaneously determining the sulfur content from the amount of SO_2 evolved. The SO_2 is trapped by a solution of iodine and its excess is titrated with thiosulfate or absorbed by 3% hydrogen peroxide, with subsequent titration of the sulfuric acid so formed against alkali.

Literature Cited

1. V. A. Obolonchik, V. F. Bukhanevich, and S. V. Radzikovskaya, Poroshkovaya Met. (1968).
2. V. Kh. Oganesyan, V. F. Bukhanevich, and S. V. Radzikovskaya, Dokl. Akad. Nauk Arm. SSR (1966).
3. G. V. Samsonov and S. V. Radzikovskaya, Uspekhi Khim., 30:61 (1961).

ANALYSIS OF THE SILICIDES OF
GROUP IV TO VI TRANSITION ELEMENTS

V. P. Kopylova and T. N. Nazarchuk

Institute of Problems in Materials Science, Academy of Sciences of the Ukrainian SSR

Present-day requirements in analytical chemistry reduce fundamentally to increasing the sensitivity, accuracy, and selectivity of methods of determining various elements. The analysis of refractory compounds is of particular interest in this connection in view of the high chemical stability of these analytically intractable materials.

The silicides of Group IV-VI transition elements are widely employed in modern technology. The analysis of these substances is complicated by the high chemical stability of the silicides. In addition to this, the decomposition of the silicides involves the evolution of gaseous decomposition products, namely, silanes of various compositions, and this may lead to inaccurate results.

Ordinary methods of decomposing the silicides (melting with soda and sodium peroxide) are not particularly satisfactory for analytical chemists, as they either lead to the loss of scarce platinum (in the case of melting with soda) or else to the possible loss of silicon in the form of silanes, which are often evolved in the course of melting with sodium peroxide; the analysis is also complicated by the fact that large quantities of iron or nickel pass into the solution. It is also sometimes recommended to decompose insoluble silicates by melting with a mixture of caustic soda and sodium peroxide in nickel crucibles [2]. A far more promising method of decomposing refractory compounds is that of baking, as already used for decomposing boron nitride and carbide, the borides of transition metals, and certain nitrides when determining the nitrogen in them [1, 3-5].

We tried using this method for decomposing the silicides of Group IV-VI transition metals. Preliminary experiments showed that the silicides of titanium, zirconium, hafnium, vanadium, niobium, tantalum, molybdenum, tungsten, and chromium, and also silicon carbide and nitride decomposed on baking with a mixture of soda and zinc oxide. The baking should be carried out in a nickel crucible at 900-1000°C. For this purpose a 0.1-g sample of the silicide is carefully mixed with 2 g of a previously-dried mixture of sodium carbonate and zinc oxide in 2:1 ratio (volumetric). At the bottom of the crucible is a "cushion" of zinc oxide (no more than 1 g in weight); the sample is covered with the mixture to be baked (about 1 g) and baked for 1.5-2 h in a muffle furnace at 900-1000°C. After cooling, the cake is poured into a beaker (if there are any remains in the crucible, these are dissolved in 1:4 sulfuric acid), and 25 ml of concentrated nitric acid are added, or else the sample is processed in an appropriate manner in order to determine what chemical reactions have taken place during the baking of the silicides.

123

On baking titanium, zirconium, and hafnium silicides, the whole of the silicide silicon is contained in the aqueous extract. The ZnO of the baking mixture is dissolved in HCl. Analysis of the residue insoluble in HCl showed that the amount of metal in the latter corresponded to the stoichiometric composition of the dioxides. For Group VI elements all the silicon and metal passes completely into the aqueous extract, or rather into the solution of sodium carbonate. The residue insoluble in the soda solution corresponds (according to analysis) to ZnO.

In discussing the processes involved in the baking of Group V silicides, we must allow for the formation of the vanadate, niobate, and tantalate of the alkali metal; these pass into the aqueous extract together with sodium silicate (for vanadium) or potassium silicate and the corresponding niobate and tantalate, while the insoluble residue dissolves in HCl and is identified as the zinc oxide employed in the baking of the silicides.

Thus baked titanium, zirconium, and hafnium silicides decompose with the formation of sodium silicate and the oxide of the corresponding metal, as in the equation

$$Me^{4+}Si_2 + 2Na_2CO_3 + 3O_2 = 2Na_2SiO_3 + 2CO_2 + MeO_2.$$

The decomposition associated with the baking of chromium, tungsten, and molybdenum silicides involves the formation of sodium silicate and the sodium salt of chromic, tungstic, or molybdic acids, respectively, in accordance with the scheme

$$2Me^{6+}Si_2 + 6Na_2CO_3 + 7O_2 = 4Na_2SiO_3 + 6CO_2 + 2Na_2Me^{6+}O_4.$$

The decomposition associated with the baking of the Group V transition element silicides involves the formation of sodium silicate and sodium vanadate (for vanadium silicide) in accordance with:

$$4VSi_2 + 10Na_2CO_3 + 13O_2 = 8Na_2SiO_3 + 10CO_2 + 4NaVO_3.$$

On decomposition of the niobium and tantalum silicides, the ortho-salts are formed in accordance with the reaction

$$4MeSi_2 + 14Na_2CO_3 + 13O_2 = 8Na_2SiO_3 + 14CO_2 + 4Na_3MeO_4.$$

In the mixture employed for baking, the zinc oxide simply serves to prevent the melting of the mixture and enables the baking to take place at a higher temperature.

In the sulfuric acid separation of silicic acid, with a single evaporation up to the point of SO_3 evolution, a certain amount of the silicic acid passes into the dissolved state. We see from the data relating to a single separation of silicic acid (with a supplementary determination of the latter in the filtrate) presented in Table 1 that the amount of silicic acid remaining in the filtrate varies from 0.2 to 2.5% for a total silicon content of 50% in the chromium silicide. For a total silicon content of 38% the losses in the acid filtrate amount to ~ 0.2-1.5% (for zirconium silicide).

According to our experimental data, silicic acid is separated to the fullest extent on evaporating its solutions with HNO_3 and bringing up to the point of SO_3 evolution

TABLE 1. Sulfuric Acid Separation of Silicic Acid in Chromium and Zirconium Silicides, %

Silicide	Silicon found		Total silicon
	in first separation	in second separation	
CrSi$_2$	49.1	1.10	50.20
	49.4	0.70	50.10
	49.2	1.10	50.30
	48.6	2.00	50.60
	48.2	1.70	49.90
	48.7	2.30	51.00
	49.7	0.30	50.00
	48.6	1.10	49.70
	48.1	1.70	49.80
	48.70	1.40	50.10
ZrSi$_2$	37.60	0.20	37.80
	36.80	1.50	38.30
	38.30	0.50	38.80
	37.90	0.60	38.50
	38.30	0.40	38.70
	38.30	0.30	38.60
	36.80	1.50	38.30
	37.1	1.10	38.20

(twice), also adding a solution of gelatin during the dissolution of the salts. In this case the silicic acid is separated almost completely from aqueous solutions in an insoluble form; it may be filtered rapidly and conveniently through a "white band" filter and is practically insoluble on washing.

In order to separate the silicic acid from the second component of the alloy (the metal) we exploited the capacity of the metal to form stable complexes with various complexing agents. Thus, in order to prevent the precipitation of the poorly-soluble basic sulfate of tervalent chromium, we recommend introducing a saturated solution of oxalic acid before evaporation; this forms fairly stable complexes with the tervalent chromium ions. In the presence of hydrogen peroxide the same oxalic acid holds niobium, tantalum, and tungsten in solution on bringing the latter up to a state of SO_3 evolution. Thus by using the baking method, with the subsequent application of various complexing agents and the conditions here developed for the complete separation of the silicic acid, we have a reasonable method of determining silicon in all the silicides of the Group IV-VI elements.

Baking Method

First 0.1 g of the silicide is carefully mixed in a glass weighing bottle with 2 g of the baking mixture and carefully transferred to a nickel crucible, at the bottom of which is a closely packed "cushion" of zinc oxide (not more than 1 g). The weighing bottle is "washed" with two portions of the mixture (not more than 0.5 g each) and the charge is covered with the same mixture at the top. The silicides are baked at 900-1000°C in a muffle furnace for 1.5-2 h. The cake is formed in the guise of a small tablet, usually coming easily away from the nickel crucible. No nickel passes from the crucible into the cake during the baking. If the cake sticks to the nickel crucible it must be cleaned off mechanically.

Determination of Silicon in Titanium, Zirconium, and

Hafnium Silicides

The cake is placed in a beaker and covered with a watch glass, 25 ml of nitric acid (sp. gr. 1.43) and 10 ml of a 30% solution of perhydrol then being added. If pieces of cake remain in the nickel crucible, this must be washed with small quantities of sulfuric acid (1:4), to a total volume of 20 ml. Then 20 ml of sulfuric acid (sp. gr. 1.84) are poured into the beaker and the solution is evaporated until SO_3 fumes appear over a period of 5-10 min. The resultant solution is cooled, 25 ml of hot nitric acid (sp. g. 1.43) are added together with 50 ml of hot water to dissolve the salts and 20 ml of sulfuric acid (1:1), and again evaporation proceeds up to the point of SO_3 evolution for at least 5-10 min. Then 100 ml of a 0.15% solution of gelatin heated to 60-70°C are poured into the cooled solution, the salts are dissolved by shaking (without heating on the electric plate) and the precipitating silicic acid is filtered through a "white band" filter. In routine work it is recommended that the silicic acid should be filtered immediately after the dissolution of the salts. The silicic acid is washed in the filter firstly with hot HCl (1:99) and then two or three times with hot water; it is calcined and roasted at 1000°C in a platinum crucible to a state of constant weight. The roasted residue is treated with 2-3 ml of hydrofluoric acid and two or three drops of sulfuric acid, evaporated to dryness, roasted, and weighed.

Determination of Silicon in Vanadium and

Molybdenum Silicides

The cake is placed in a beaker, covered with a watch glass, and treated with 25 ml of nitric acid (sp.gr. 1.43). The remains of the cake are transferred to the beaker from the nickel crucible by washing with a solution of sulfuric acid (1:4) 20 ml in volume; then 20 ml of sulfuric acid (sp.gr. 1.84) are added and the solution is evaporated until SO₃ fumes appear for 5-10 min. The resultant solution is cooled, 25 ml of hot nitric acid (sp.gr. 1.43) are added together with 50 ml of hot water to dissolve the salts, followed by 20 ml of sulfuric acid (1:1), and the result is again evaporated until SO₃ appears for not less than 5-10 min. Then 100 ml of a 0.15% gelatin solution heated to 60-70°C are added to the cooled solution, the salts are dissolved by shaking (without heating on the electric plate), and the precipitating silicic acid is filtered through a "white band" filter. The precipitate is washed in the filter first with hot HCl (1:99) and then two or three times with hot water, calcined, and roasted at 1000°C in a platinum crucible to a state of constant weight. The roasted residue is treated with 2-3 ml of HF, with two or three drops of sulfuric acid, evaporated to dryness, roasted, and weighed.

Determination of Silicon in Chromium Silicide

The cake is placed in a beaker, covered with a watch glass, and 25 ml of nitric acid (sp.gr. 1.43) are added. The remains of the cake are transferred to the beaker from the nickel crucible by washing with a solution of sulfuric acid (1:4) with a total volume of 20 ml; then 20 ml of sulfuric acid (sp.gr. 1.84) and 50 ml of a saturated solution of oxalic acid are added. The resultant solution is evaporated until SO₃ fumes appear for not less than 5-10 min. After cooling, 25 ml of hot HNO₃ (sp.gr. 1.43) are added to the solution with 50 ml of hot water to dissolve the salts; 20 ml of sulfuric acid (1:1) and 50 ml of a saturated oxalic acid solution follow, and the result is again evaporated until SO₃ fumes appear for not less than 5-10 min. Then 100 ml of a 0.15% gelatin solution heated to 60-70°C are added to the cooled solution, the salts are dissolved by shaking (without heating on the electric plate), and the silicic acid is filtered through a "white band" filter. Subsequent operations are as indicated earlier.

Determination of Silicon in Niobium, Tantalum, and

Tungsten Silicides

The cake is placed in a beaker and covered with a watch glass; 25 ml of nitric acid (sp. gr. 1.43) are added, with 10-15 ml of a 30% solution of perhydrol, 20 ml of sulfuric acid (sp.gr. 1.84), and 100 ml of a saturated solution of oxalic acid, and the solution is evaporated until the niobic, tantalic, or tungstic acids are completely dissolved. After cooling, 25 ml of hot nitric acid (sp.gr. 1.43) are added with 10 ml of a 30% solution of perhydrol, 100 ml of a heated saturated solution of oxalic acid, and 20 ml of sulfuric acid (1:1), and the result is evaporated again until SO₃ fumes appear and the niobic, tantalic, or tungstic acids are completely dissolved. To the cooled solution we add 10 ml of a 30% solution of perhydrol, shake, add 100 ml of a 0.15% solution of gelatin in a saturated solution of oxalic acid heated to 60-70°C, dissolve the salts by shaking (without heating on the electric plate), and filter the silicic acid on a "white band" filter. The residue is washed on the filter with hot HCl (1:99) and then two or three times with hot water, calcined, and roasted to constant weight at 1000°C in a platinum crucible. The roasted residue is treated with 2-3 ml of HF and two or three drops of sulfuric acid, evaporated to dryness, roasted, and weighed.

The foregoing method was employed in order to analyze a number of transition metal silicides. For comparison, some of these were also analyzed by existing methods, involving decomposition by melting with soda in platinum crucibles and twice precipitating the silicic acid, melting with sodium peroxide in nickel crucibles and twice precipitating the silicic acid, and baking with a mixture of soda and ZnO (Table 2). The results presented show that the proposed method is in no way inferior to existing methods as regards accuracy of determination, yet it proves economical in respect of time as it only requires one precipitation of silicic acid.

TABLE 2. Determination of Silicon in Silicides, %

Silicide	Melting with sodium per-oxide and double pre-cipitation of silicic acid	Melting with soda and dou-ble precipi-tation of silicic acid	Baking with mixture of soda and ZnO and a single precipitation of silicic acid	Deviation from mean value, %	Mean square error of the method
HfSi$_2$	25.2	25.1	25.0 25.2 24.8 25.5	—0.1 +0.1 —0.3 +0.4	$2.25 \cdot 10^{-2}$
ZrSi$_2$	34.4	34.2	34.4 34.5 34.3	+0.1 —0.1 0	$0.33 \cdot 10^{-2}$
VSi$_2$	48.6	48.5	48.7 48.9 48.3	+0.1 +0.3 —0.3	$3.16 \cdot 10^{-2}$
CrSi$_2$	48.6	48.4	48.5 48.8 48.6	—0.1 +0.2 0	$0.83 \cdot 10^{-2}$

Conclusions

1. In order to decompose the silicides of Group IV-VI transition metals, we have proposed a method of baking with a mixture of zinc oxide and anhydrous sodium carbonate in nickel crucibles.

2. We have developed a reasonable and efficient method of determining the total silicon content in the silicides of Group IV-VI transition metals by the sulfuric acid process, involving a single precipitation of silicic acid.

Literature Cited

1. Analysis of Refractory Compounds [in Russian], Gos. Nauch.-Tekh. Izd. Lit. Chern. i Tsvet. Met., Moscow (1962).
2. Inform. Quim. Analit., 14(2):38-40 (1960).
3. K. D. Modylevskaya, M. D. Lyutaya, and T. N. Nazarchuk, Zavod. Lab., 27:1345 (1961).
4. K. D. Modylevskaya, M. D. Lyutaya, and T. N. Nazarchuk, Zh. Neorg. Khim., 4:12 (1961).
5. N. N. Minenko and T. N. Nazarchuk, Poroshkovaya Met., No. 6 (1965).